T0298680

Chemistry of the f-Block Elements

ADVANCED CHEMISTRY TEXTS

A series edited by DAVID PHILLIPS, *Imperial College, London, UK*, PAUL O'BRIEN, *University of Manchester, UK* and STAN ROBERTS, *University of Liverpool, UK*

Volume 1
Chemical Aspects of Photodynamic Therapy
Raymond Bonnett

Volume 2
Transition Metal Carbonyl Cluster Chemistry
Paul J. Dyson and J. Scott McIndoe

Volume 3
Nucleoside Mimetics: Their Chemistry and Biological Properties
Claire Simons

Volume 4
Sol-Gel Materials: Chemistry and Applications
John D. Wright and Nico A.J.M. Sommerdijk

Volume 5
Chemistry of the f-Block Elements
Helen C. Aspinall

This book is part of a series. The publisher will accept continuation orders which may be cancelled at any time and which provide for automatic billing and shipping of each title in the series upon publication. Please write for details.

Chemistry of the f-Block Elements

Helen C. Aspinall

University of Liverpool, UK

Gordon and Breach Science Publishers

Australia • Canada • France • Germany • India • Japan
Luxembourg • Malaysia • The Netherlands • Russia
Singapore • Switzerland

 Printed in Singapore.

Amsteldijk 166
1st Floor
1079 LH Amsterdam
The Netherlands

British Library Cataloguing in Publication Data

Aspinall, Helen C.
Chemistry of the f-block elements. – (Advanced chemistry texts ; v. 5)
1. Rare earth metals 2. Actinide elements
I. Title
546.4′1

ISBN: 90-5699-333-X
ISSN: 1027-3654

In memory of Mum

Contents

Preface

This book is intended to fill a gap between the rather brief accounts of the f-block elements given in textbooks on inorganic chemistry, and the very detailed accounts available in specialized monographs. The content is based on a lecture course which is given to final year undergraduates at Liverpool, and has been selected to illustrate the main features of f-block chemistry, along with some of the important applications. I have necessarily omitted a great many compounds, but accounts of these can be found elsewhere. The chemistry of the lanthanides and actinides has expanded dramatically in the last decade, and I have included many references to the original literature which I hope the reader will find useful in gaining access to recent developments in the area.

I am grateful to Professor D. C. Bradley FRS who introduced me to lanthanide chemistry when I was a postdoctoral researcher at Queen Mary College London, and to research students at Liverpool, whose enthusiasm has contributed greatly to my continued (and growing) interest in the area. Almost all of the structural diagrams have been prepared using data from the Cambridge Crystallographic Database or the Inorganic Crystal Structure Database, accessed via the Chemical Database Service at Daresbury. I am grateful to colleagues for supplying me with other diagrams; they are acknowledged at the appropriate points in the text. Lastly, a very big 'thank you' to my husband, Dom, who has been a tremendous support (as always) while I have been working on this book.

CHAPTER 1

INTRODUCTION

The lanthanide elements are usually defined as those in which the 4f-orbitals are progressively filled; this definition includes the elements Ce (cerium) to Lu (lutetium). Although not a lanthanide by this definition, La (lanthanum) is the prototype for the series and for practical purposes is usually included. The term 'rare earth' dates back to the early 19th century and is applied to elements 57 to 71 (La to Lu) and also to Y (yttrium), which is found in nature along with the lanthanides. Sc (scandium) is often also classed as a rare earth. The elements Th (thorium) to Lr (lawrencium) form the actinide series which is formally a 5f analogue of the lanthanide series, with Ac (actinium) as its prototype. This 'actinide concept', proposed in 1944 by Glenn Seaborg, solved the problem of where to fit the trans-uranium elements into the periodic table. The chemistry of the early actinides is quite distinct from that of their 4f congeners, and prior to 1944 the elements Th to U had been placed in the periodic table immediately below Hf, Ta and W. The late actinides show many similarities to the lanthanides.

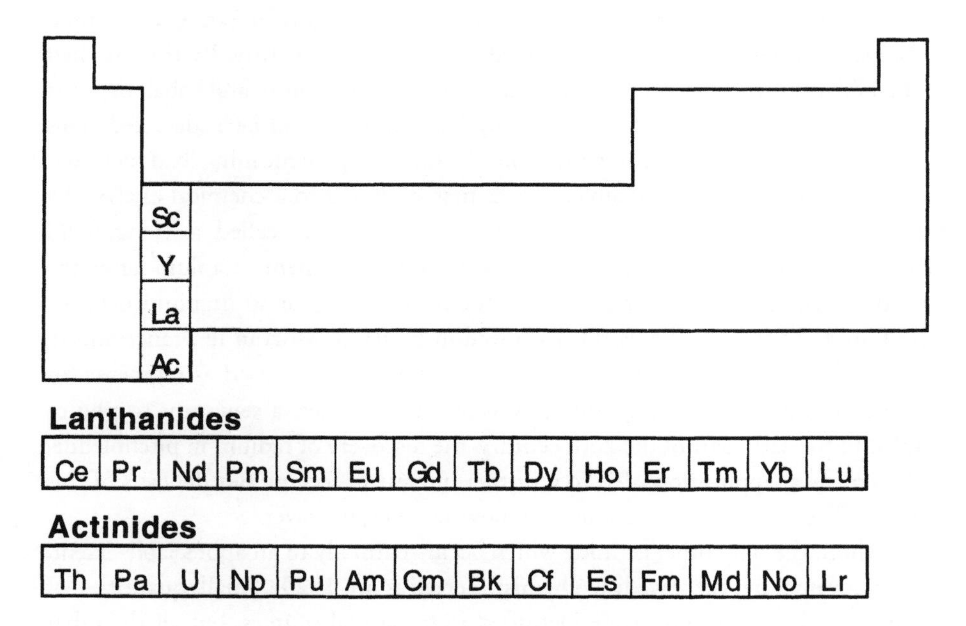

FIGURE 1.1 THE RARE EARTH AND ACTINIDE ELEMENTS IN THE PERIODIC TABLE.

1.1 HISTORY

1.1.1 Lanthanides

The story of the lanthanides begins in 1787 when a young Swedish artillery officer, Lieutenant Carl Axel Arrhenius, who was a keen amateur geologist, was exploring a quarry at a small town called Ytterby , near Stockholm. He found a new, very dense black mineral which he named 'ytterbite'. At the time there was some speculation that the mineral might contain the recently discovered element tungsten, but the first serious chemical analysis was carried out in 1794 by Johan Gadolin, a Finnish chemist. Methods of chemical analysis were limited in the 18th century, but after a series of treatments with acids and alkalis, Gadolin was able to show that the new mineral contained oxides of iron, beryllium, and silicon and a new, previously unidentified 'earth' which he named 'yttria'. (At the time, the term 'earth' was applied rather loosely to insoluble metal oxides.) Yttria was later shown to be a mixture of the oxides of six rare earth elements. A great deal of painstaking work was to follow during the 19th century: the first pure sample of dysprosium was obtained after 58 recrystallisations, and the first sample of Tm took 11000 crystallisations! With the discovery of lutetium (Lu) in 1907, the naturally occurring rare earths had all been isolated. Mosely's pioneering work on X-ray spectroscopy in the early 20th century was invaluable in determining the purity (or otherwise) of newly discovered elements and also in pointing out the gaps in the Periodic Table. The missing element number 61, promethium, was synthesised and characterised in 1947, completing the lanthanide series. For an account of the history of promethium see Cotton (1999). Some of the history of the lanthanides is summarized in Table 1.1.

1.1.2 Actinides

Uranium was the first of the f-block elements to be discovered: its history dates back to 1789, the year of the French Revolution. Its story begins in Sankt Joachimstal in Bohemia, where silver had been mined since the 16th century. By the first half of the 17th century, silver mining had almost ceased, but bismuth and cobalt deposits were still being exploited, and a new shiny black mineral had been detected. This mineral was nicknamed 'pitchblende' from the German *pech* meaning 'bad luck' and *blende* meaning 'mineral'. Pitchblende was first subjected to a chemical analysis by the German chemist Martin Klaproth who isolated what he called 'a strange kind of half metal' from the mineral. He named the new element 'uranium' after the recently discovered planet Uranus. Over the next century or so uranium deposits were found throughout the world: Cornwall in England, Morvan in France and in Austria and Romania. Uranium oxides and salts were widely used as pigments for ceramics and glass, and uranyl nitrate was also used to give a sepia tint to photographs. In the early part of the 20th century, the discovery of radium in pitchblende, and its medical applications, led to a further interest in seeking out deposits of this mineral. The largest use of uranium is now in nuclear power.

Thorium, the only other actinide which occurs naturally to an appreciable extent, was discovered by the Swedish chemist Berzelius in 1829. Small quantities of actinium and protactinium were identified from natural sources, but all the other actinides (from neptunium to lawrencium) were synthesized between 1940 and 1961

TABLE 1.1 HISTORY OF THE LANTHANIDES.

Element	Discoverer	Date	Origin of name
Cerium, Ce	C.G. Mosander	1839	Asteroid, *Ceres*
Lanthanum, La	C.G. Mosander	1839	*lanthanein*: to escape notice
Praseodymium, Pr	C.A. von Welsbach	1885	*praseos+didymos*: green twin
Neodymium, Nd	C.A. von Welsbach	1885	*neos+didymos*: new twin
Samarium, Sm	L. de Boisbaudran	1879	mineral *samarskite*
Europium, Eu	E.A. Demarcay	1901	*Europe*
Yttrium, Y	C.G. Mosander	1843	*Ytterby*
Terbium, Tb	C.G. Mosander	1843	*Ytterby*
Erbium, Er	C.G. Mosander	1843	*Ytterby*
Ytterbium, Yb	J.C.G. de Marignac	1878	*Ytterby*
Scandium, Sc	L.F. Nilson	1879	*Scandinavia*
Holmium, Ho	P.T. Cleve	1879	Stock*holm*
Thulium, Tm	P.T. Cleve	1879	*Thule*: most northerly land
Gadolinium, Gd	J.C.G. de Marignac	1880	Finnish chemist J *Gadolin*
Dysprosium, Dy	L. de Boisbaudran	1886	*dysprositos*: hard to get
Lutetium, Lu	G. Urbain C.A. von Welsbach C. James	1907	*Lutetia:* Paris
Promethium, Pm	J.A. Marinsky L.E. Glendenin C.D. Coryell	1947	*Promethius*: the mythological giver of fire to mankind

following the invention of the cyclotron by E. O. Lawrence in 1930. Most of this work was carried out by Glenn Seaborg and his team at Berkeley in California. Some of the history of the actinides is summarized in Table 1.2.

1.2 SOURCES OF THE ELEMENTS

1.2.1 Lanthanum, yttrium and the lanthanides

Despite being known as the 'rare earths', lanthanum, yttrium and the lanthanides are by no means rare: Tm, the least abundant naturally occurring lanthanide, is more abundant than mercury or iodine. The only lanthanide which does not occur naturally is promethium, Pm, which was prepared in 1947 by neutron irradiation of Nd. The most important mineral source of the rare earths is bastnasite, which is a mixed lanthanide fluorocarbonate of formula $LnFCO_3$. The rare earth content of bastnasite is mainly Ce (49%), La (33%), Nd (12%) and Pr (4%), with the later lanthanides being present in significantly smaller amounts. The other major rare earth ore is monazite, a mixed rare earth/thorium phosphate. This dense mineral often occurs as a sand, and large deposits are found on the Indian Ocean coast of Australia, in southern India and in China and Brazil. Because of its high Th content

TABLE 1.2 HISTORY OF THE ACTINIDES.

Element	Symbol	Discoverer	Date	Origin of name
Uranium	U	M.H. Klaproth	1789	Recently discovered planet *Uranus*
Thorium	Th	J.J. Berzelius	1828	*Thor*, Scandinavian god of war
Actinium	Ac	A. Debierne	1899	Radio*activity* of ^{227}Ac & its decay products
Protactinium	^{234}Pa	K. Fajans and O. Gohring	1913	Originally called Brevium due to short $t_{1/2}$ (6.66h)
	^{231}Pa	O. Hahn & L. Meitner, F. Soddy & J.A. Cranston	1916	*Protoactinium*, decays to form ^{227}Ac
Neptunium	^{239}Np	E.M. McMillan & P.H. Abelson	1940	*Neptune*, the planet beyond Uranus
Plutonium	Pu	G.T. Seaborg, E.M. McMillan, J.W. Kennedy & A. Wahl	1940	*Pluto*, the planet beyond Neptune
Americium	Am	G.T. Seaborg, R.A. James, L.O. Morgan & A. Ghiorso	1944	*America*, by analogy with *Europium*
Curium	Cu	G.T. Seaborg, R.A. James, & A. Ghiorso	1944	P. and M. *Curie*, by analogy with *Gadolinium*
Berkelium	Bk	S.G. Thompson, A. Ghiorso and G.T. Seaborg	1949	*Berkeley*, by analogy with Tb, named after *Ytterby*
Californium	Cf	S.G. Thompson, K. Street, A. Ghiorso and G.T. Seaborg	1950	*California*, location of the lab
Einsteinium	Es	Workers at Berkeley, Los Alamos and Argonne	1952	Albert *Einstein*
Fermium	Fm	Workers at Berkeley, Los Alamos and Argonne	1952	Enrico *Fermi*, inventor of the first self-sustaining nuclear reactor
Mendelevium	Md	A. Ghiorso, B.H. Harvey, G.R. Choppin, S.G. Thompson & G.T. Seaborg	1955	Dimitri *Mendeleev*
Nobelium	No	A. Ghiorso, T. Sikkeland, J.R. Walton & G.T. Seaborg	1958	Alfred *Nobel*, benefactor of science
Lawrencium	Lr	A. Ghiorso, T. Sikkeland, A.E. Larsch & R.M. Latimer	1961	Ernest *Lawrence*, developer of the cyclotron

TABLE 1.3 1996 RARE EARTH PRODUCTION (TONNES PER ANNUM)

Country	Tonnes
China	55000
USA	20400
USSR	6000
India	2700
Others	665
Total	84765

(approximately 10%) and consequent radioactivity, safety considerations are making monazite a much less attractive rare earth source.

The initial processing of monazite is usually by reaction with NaOH at about 150°C for several hours. This results in formation of insoluble $Ln(OH)_3$ and $Th(OH)_4$, which can easily be separated, along with soluble Na_3PO_4. The solids are then reacted with HCl at approximately pH 3–4. Under these conditions the $Th(OH)_4$ is unreactive and the $Ln(OH)_3$ dissolve to give an aqueous solution of $LnCl_3$. Bastnasite may be leached directly with acid to give an aqueous solution of all the Ln, or it may be roasted in air to first oxidize Ce(III) to Ce(IV). Ce(IV) will not be extracted in a leach with dilute acid and so can be separated from the other Ln.

Further separation of the elements is necessary for many of the applications of the rare earths, and it is carried out industrially by a solvent extraction process which is described in Chapter 3. The Ln metals can be obtained by reduction of fluorides, chlorides or oxides with an active metal such as Ca or Li, or electrolytically as a solution in a molten salt of an alkali or alkaline earth metal.

Total production of rare earth oxides in 1996 was approximately 85000 tonnes; the production by country is summarized in Table 1.3.

1.2.2 Actinides

Only two actinides, thorium and uranium, occur naturally to any significant extent, and both of these elements are reasonably abundant. The main source of thorium is monazite, where it is found along with the rare earths. Actinium and protactinium are decay products of the naturally occurring ^{235}U and so are present in very small quantities in uranium ores. Approximately 130 g of protactinium was obtained by repeated extraction of 60 tonnes of pitchblende residues in the 1960's.

$$^{235}_{92}U \rightarrow ^{231}_{90}Th + \alpha \rightarrow ^{231}_{91}Pa + \beta^- \rightarrow ^{227}_{89}Ac + \alpha$$

Pitchblende or uraninite, U_3O_8, is the most important ore for uranium. In the first stage of processing, the ore is ground to a powder and then treated either with sulfuric acid to give a solution of uranyl sulfate, or with a mixture of Na_2CO_3 and

TABLE 1.4 1998 URANIUM PRODUCTION (TONNES PER ANNUM)

Country	Tonnes
Canada	10924
Australia	4885
Niger	3731
Namibia	2762
Russia	2000
Uzbekistan	2000
USA	1872
Kazakhstan	1250
South Africa	962
Others	3546
Total	33932

$NaHCO_3$ to give a solution of uranyl carbonate. The aqueous solutions are then separated from insoluble materials and are further purified by solvent extraction (see Chapter 3) which separates uranyl ion UO_2^{2+} from other dissolved species. The product of the solvent extraction process is pure uranyl nitrate $UO_2(NO_3)_2$. This can be converted thermally to UO_3, and then by H_2 reduction to UO_2 which is the fuel in modern nuclear reactors. Uranium metal can be prepared by reduction of UF_4 with Mg. Total world uranium production in 1998 was approximately 34 000 tonnes; the major producers are summarized in Table 1.4.

The main source of actinides other than Th and U is neutron irradiation of lighter elements in nuclear reactors as shown in the schemes below:

$$^{226}_{88}Ra + ^{1}_{0}n \rightarrow ^{227}_{88}Ra + \gamma \xrightarrow{-\beta^-} ^{227}_{89}Ac$$

$$^{230}_{90}Th + ^{1}_{0}n \rightarrow ^{231}_{90}Th + \gamma \xrightarrow{-\beta^-} ^{231}_{91}Pa$$

$$^{238}_{92}U + ^{1}_{0}n \rightarrow ^{237}_{92}U + 2^{1}_{0}n$$

$$^{237}_{92}U \rightarrow ^{237}_{93}Np + \beta^-$$

$$^{235}_{92}U + ^{1}_{0}n \rightarrow ^{236}_{92}U + \gamma \xrightarrow{^{1}_{0}n} ^{237}_{92}U + \gamma \xrightarrow{-\beta^-} ^{237}Np$$

$$^{235}_{92}U + ^{1}_{0}n \rightarrow ^{239}_{92}U \xrightarrow{-\beta^-} ^{239}_{93}Np \xrightarrow{-\beta^-} ^{239}_{94}Pu$$

$$^{237}_{93}Np + ^{1}_{0}n \rightarrow ^{238}_{93}Np \xrightarrow{-\beta^-} ^{237}_{94}Pu$$

TABLE 1.5 LONGEST-LIVED ISOTOPES OF THE ACTINIDES.

Isotope	Half-life	Decay mode[a]
^{227}Ac*	21.77 y	β(0.0.41), α (5.043)
^{232}Th*	1.41×10^{10} y	α (4.081)
^{231}Pa*	3.27×10^4 y	α (5.148)
^{234}U*	2.45×10^5 y	α (4.856)
^{235}U*	7.04×10^8 y	α (4.6793)
^{238}U*	4.46×10^9 y	α (4.039)
^{237}Np	2.14×10^6 y	α (4.957)
^{238}Pu	87.74 y	α (5.593)
^{239}Pu	2.411×10^4 y	α (5.244)
^{240}Pu	6.57×10^3 y	α (5.255)
^{242}Pu	3.76×10^5 y	α (4.983)
^{244}Pu	8.2×10^7 y	α (4.665), SF
^{241}Am	433 y	α (5.637)
^{243}Am	7.37×10^3 y	α (5.439)
^{244}Cm	18.1 y	α (5.902)
^{245}Cm	8.5×10^3 y	α (5.623)
^{246}Cm	4.78×10^3 y	α (5.476)
^{247}Cm	1.6×10^7 y	α (5.352)
^{248}Cm	3.4×10^5 y	α (5.162), SF
^{247}Bk	1.4×10^3 y	α (5.889)
^{249}Bk	320 d	β
^{249}Cf	351 y	α (6.295)
^{251}Cf	890 y	α (6.172)
^{252}Cf	2.64 y	α (6.217), SF
^{252}Es	1.29 y	α (6.739)
^{254}Es	275 d	α (6.617)
^{255}Es	39.8 d	β
^{257}Fm	100.5 d	α (6.871)
^{258}Md	56 d	α (6.716) 72%, α (6.79) 28%, SF
^{259}No	58 min	α (7.794) 78%, EC 22%
^{260}Lr	3 min	α (8.30)

[a]Energy of radiation (in MeV) in parentheses; SF = spontaneous fission, EC = electron capture
*Naturally occurring isotope

The elements americium to fermium are usually obtained as by-products of the large-scale production of ^{239}Pu. Formation of trans-plutonium elements by slow neutron capture requires very high neutron fluxes and often prolonged times: for example irradiation of 1 kg plutonium for 5–10 years at a neutron flux of 3×10^{14} cm^{-2} s^{-1} would produce approximately 1 mg of californium. Several of these elements are only available in mg quantities or less per year, and investigation of their chemistry is only made possible by the use of tracer techniques. The longest-lived isotopes of the actinides are summarized in Table 1.5.

1.3 APPLICATIONS OF RARE EARTHS AND ACTINIDES

1.3.1 Lanthanum, yttrium and the lanthanides

The first commercial application of a rare earth element was made by the Austrian

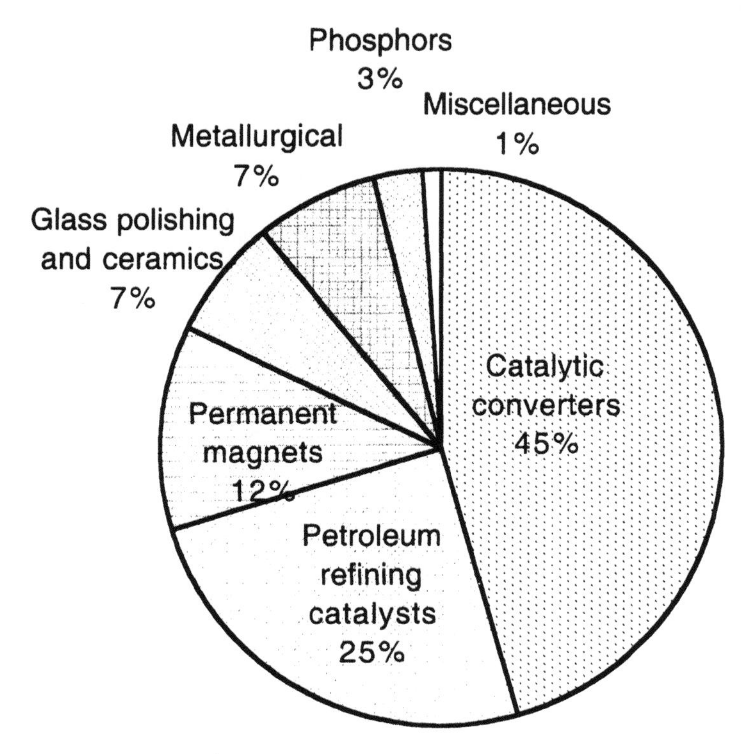

FIGURE 1.2 THE APPLICATIONS OF RARE EARTHS (1996).

chemist and entrepreneur Carl Auer von Welsbach. In 1891 he patented a gas mantle which consisted of 99% ThO_2 and 1% CeO_2. This mantle made possible the widespread use of incandescent gas lamps both indoors and outdoors. Von Welsbach was also responsible for another early rare-earth application: a means of lighting the gas lamp. The use of all the Th and a small quantity of Ce from the monazite ore left him with large quantities of mixed Ln. During his early work with Robert Bunsen on electrolytic reduction of rare earth chlorides using Fe electrodes he had observed that the alloy formed from Fe and mixed Ln was pyrophoric, emitting sparks when scratched. This 'pyrophoric alloy', consisting of approximately 30% Fe and 70% mixed rare earth metals, is often known as 'mischmetall' and is used for lighter flints. The applications of rare earths (1996 figures) are summarized in Figure 1.2.

The largest volume application of rare earths is the use of CeO_2 in catalytic converters where it acts both as a ceramic support for the platinum metal components of the catalyst, and as an oxygen reservoir due to its ability to exist as Ce_2O_3 under reducing conditions and as CeO_2 under oxidizing conditions. The use of rare earths in catalysts for petroleum cracking is also important: Ln ions can stabilize zeolite structures under the operating conditions of the catalyst. The introduction of $SmCo_5$ and Sm_2Co_{17} as permanent magnet materials led to a large increase in demand for samarium, now accounting for approximately 12% of world consumption. Some of the smallest volume applications are also among the highest value, and the special electronic and magnetic properties of rare earth ions are exploited in high value applications. These include Eu phosphors (which give the red colour in TV screens), 'Colour 80' fluorescent lamps, Nd-YAG lasers, and contrast agents for magnetic resonance imaging (used in diagnostic medicine), all of which are described in Chapter 2. Rare earth sulfides are used as non-toxic alternatives to Cd in pigments for plastics; the colours obtainable range from yellow through orange to burgundy. The unique reactivity of lanthanide complexes is being exploited in their use as catalysts and reagents in synthetic chemistry, which is described in Chapter 5.

1.3.2 Actinides

Thorium

The first application of thorium was as the major component in gas mantles at the end of the 19th century. More recently there have been proposals to use thorium as a source of nuclear fuel. Although ^{232}Th is non fissile, it can capture a neutron and ultimately form ^{233}U which is fissile.

$$^{232}_{90}\text{Th}+^{1}_{0}\text{n}\rightarrow^{233}_{90}\text{Th}\rightarrow^{233}_{91}\text{Pa}+\beta^- \quad 23\text{ minutes}$$

$$^{233}_{91}\text{Pa}\rightarrow^{233}_{92}\text{U}+\beta^- \quad 27\text{ days}$$

Current world production of thorium (1998) is approximately 1500 tonnes per annum. The major application (>50%) is in refractory materials; other applications are in ceramics, lighting and aerospace alloys.

Uranium

Uranium(IV) oxide has been used for many years as a yellow pigment for glass and ceramics, but apart from this relatively small scale application, the main use of uranium is as a nuclear fuel. Modern nuclear reactors use UO_2 as a fuel, and for this application the concentration of the fissile ^{235}U must be increased to between 3 and 5% as described in Section 1.5.2.

$$^{235}_{92}\text{U}+^{1}_{0}\text{n}\rightarrow^{90}_{36}\text{Kr}+^{142}_{56}\text{Ba}+2-3^{1}_{0}\text{n}$$

Fissile ^{239}Pu is also produced during this nuclear reaction by addition of a neutron to ^{238}U.

Plutonium

^{238}Pu is used in long-lasting portable power sources (for example those used in space probes) and in the miniature batteries used in heart pacemakers. These radioisotope thermoelectric generators (or RTG's) rely on the heat produced during the α-decay of ^{238}Pu. This heat is converted into electricity by a silicon-germanium thermopile inside the RTG. The Galileo probe, which was launched in 1989 to investigate the planet Jupiter, used a 50W RTG containing about 22kg of PuO_2; a typical heart pacemaker power supply contains about 160 mg of PuO_2.

Americium

^{241}Am is an α and γ emitter with a half-life of 433 years and is found in most homes as a component of domestic smoke detectors. The price of AmO_2 in 1997 was $1500 per g, virtually unchanged since 1962; 1g of AmO_2 is sufficient for 5000 smoke detectors. When smoke particles enter the ionization chamber of the smoke detector, they absorb the α-particles, resulting in a fall in electric current, which sets off the alarm. This is the major use of the several kg of AmO_2 produced each year.

1.4 PROPERTIES OF THE ELEMENTS

Some of the most important properties of the lanthanides and actinides are summarized in Tables 1.6 and 1.7.

1.4.1 The metals

All of f-block elements are very electropositive and tarnish rapidly in air. The lanthanide metals generally show hexagonal close packed or cubic close packed structures at room temperature, and body centred cubic structures at high temperatures. The structures of the actinide metals are much more complex, especially for the early part of the series where Pu has six allotropic modifications, and U and Np have three. The elements Th to Pu all adopt a bcc structure at the melting point; Am to Es adopt a fcc structure at the melting point and double hexagonal close packed at lower temperatures. The complexity of the structures of the early actinides is due to the increased radial extent of the 5f orbitals in these elements, which allows them to affect interatomic interactions in the solid state.

1.4.2 Shapes of f-orbitals

The seven f-orbitals have m_l values of 0, ±1, ±2 and ±3. Of these, the orbital with $m_l=0$ is real, but those with $m_l=\pm1$, ±2 and ±3 are complex, and in order to obtain real orbitals, linear combinations must be taken. If simple linear combination of the $m_l=\pm1$, ±2 and ±3 pairs are taken, the resulting Cartesian functions (with abbreviated labels in brackets) are: $z(2z^2-3x^2-3y^2)$ [z^3]; $x(4z^2-x^2-y^2)$ [xz^2]; $y(4z^2-x^2-y^2)$ [yz^2]; xyz; $z(x^2-y^2)$; $x(x^2-3y^2)$; $y(3x^2-y^2)$. This set of orbitals is referred to as the 'general set' and is illustrated in Figure 1.3. For systems of cubic symmetry it can be useful to take a different set of linear combinations referred to as the 'cubic set'. The abbreviated Cartesian labels for the cubic set of orbitals, and their symmetry in O_h, are: xyz (A_{2u}); x^3, y^3, z^3 (T_{1u}); $z(x^2-y^2)$, $x(z^2-y^2)$, $y(z^2-x^2)$ (T_{2u}).

1.4.3 The lanthanide and actinide contractions

There is a steady decrease in metallic and ionic radii of the lanthanides as the series is traversed from La to Lu shown in Figure 1.4. A similar contraction is found for the actinide series (see Figure 1.5) although because of practical difficulties the data are not so complete for the actinides. (The ionic radii are Shannon's 6-coordinate radii.) The so-called 'lanthanide contraction' is responsible for the close similarity in radii of 2nd and 3rd row transition metals belonging to the same group, and it has been explained as an electrostatic effect due to the imperfect screening by the f-electrons as the nuclear charge is increased. However, more recent relativistic calculations by Laerdahl *et al.* (1988) have shown that between 10% and 30% of

TABLE 1.6 PROPERTIES OF THE LANTHANIDES.

Symbol	Name	Terrestrial abundance (ppm)	Electron configuration (metal)	Electron configuration (Ln^{3+})	Metallic radius (pm)	Ln^{3+} radius (pm)	E° (Ln^{3+}/Ln) (V)	1st IP ($kJmol^{-1}$)	2nd IP ($kJmol^{-1}$)	3rd IP ($kJmol^{-1}$)	4th IP ($kJmol^{-1}$)
Y	Yttrium	30	$[Kr]4d^1 5s^2$	[Kr]	181	104	−2.372	616	1181	1980	5963
La	Lanthanum	32	$[Xe]5d^1 6s^2$	[Xe]	187.7	117.2	−2.522	538.1	1067	1850	4819
Ce	Cerium	68	$[Xe]4f^1 5d^1 6s^2$	$[Xe]4f^1$	182.5	115	−2.483	527.4	1047	1949	3547
Pr	Praseodymium	9.5	$[Xe]4f^3 6s^2$	$[Xe]4f^2$	182.8	113	−2.462	523.1	1018	2086	3761
Nd	Neodymium	38	$[Xe]4f^4 6s^2$	$[Xe]4f^3$	182.1	112.3	−2.431	529.6	1035	2130	3899
Pm	Promethium	4.5×10^{-20}	$[Xe]4f^5 6s^2$	$[Xe]4f^4$	181	111	−2.423	535.9	1052	2150	3970
Sm	Samarium	7.9	$[Xe]4f^6 6s^2$	$[Xe]4f^5$	180.2	109.8	−2.414	543.3	1068	2260	3990
Eu	Europium	2.1	$[Xe]4f^7 6s^2$	$[Xe]4f^6$	204.2	108.7	−2.407	546.7	1085	2404	4110
Gd	Gadolinium	7.7	$[Xe]4f^7 5d^1 6s^2$	$[Xe]4f^7$	180.2	107.8	−2.397	592.5	1167	1990	4250
Tb	Terbium	1.1	$[Xe]4f^9 6s^2$	$[Xe]4f^8$	178.2	106.3	−2.391	564.6	1112	2114	3839
Dy	Dysprosium	6	$[Xe]4f^{10} 6s^2$	$[Xe]4f^9$	177.3	105.2	−2.353	571.9	1126	2200	4001
Ho	Holmium	1.4	$[Xe]4f^{11} 6s^2$	$[Xe]4f^{10}$	176.6	104.1	−2.319	580.7	1139	2204	4100
Er	Erbium	3.8	$[Xe]4f^{12} 6s^2$	$[Xe]4f^{11}$	175.7	103	−2.296	588.7	1151	2194	4115
Tm	Thulium	0.48	$[Xe]4f^{13} 6s^2$	$[Xe]4f^{12}$	174.6	102	−2.278	596.7	1163	2285	4119
Yb	Ytterbium	3.3	$[Xe]4f^{14} 6s^2$	$[Xe]4f^{13}$	194	100.8	−2.267	603.4	1176	2415	4220
Lu	Lutetium	0.51	$[Xe]4f^{14} 5d^1 6s^2$	$[Xe]4f^{14}$	173.4	100.1	−2.255	523.5	1340	2022	4360

Table 1.7 Properties of the actinides.

Symbol	Name	Electron configuration (metal)	Metallic radius (pm)	An^{3+} radius (pm)	An^{4+} radius (pm)	E° (An^{3+}/An) (V)	E° (An^{4+}/An) (V)
Ac	Actinium	[Rn]$6d7s^2$	187.8	126		−2.13	
Th	Thorium	[Rn]$6d^27s^2$	179.8 (α) 180 (β)		108		−1.83
Pa	Protactinium	[Rn]$5f^26d7s^2$	164.2 (α) 177.5 (β)	118	104		−1.47
U	Uranium	[Rn]$5f^36d7s^2$	154.2 (α) 154.8 (β)	116	103	−1.66	--1.38
Np	Neptunium	[Rn]$5f^46d7s^2$	150.3 (α) 151.1 (β)	115	101	−1.79	−1.30
Pu	Plutonium	[Rn]$5f^67s^2$	152.3 (α) 157.1 (β)	114	100	−2.00	−1.25
Am	Americium	[Rn]$5f^77s^2$	173 (α, β)	111.5	99	−2.07	−0.90
Cm	Curium	[Rn]$5f^76d7s^2$	174.3 (α) 178.2 (β)	111	99	−2.06	
Bk	Berkelium	[Rn]$5f^97s^2$	170.4 (α) 176.7 (β)	110	97	−1.96	
Cf	Californium	[Rn]$5f^{10}7s^2$	169.4 (α) 203.0 (β)	109	96.1	−1.91	
Es	Einsteinium	[Rn]$5f^{11}7s^2$	203 (β)			−1.98	
Fm	Fermium	[Rn]$5f^{12}7s^2$				−2.07	
Md	Mendelevium	[Rn]($5f^{13}7s^2$)				−1.74	
No	Nobelium	[Rn]($5f^{14}7s^2$)				−1.26	
Lr	Lawrencium	[Rn]($5f^{14}6d7s^2$ or $5f^{14}7s^27p$)				−2.1	

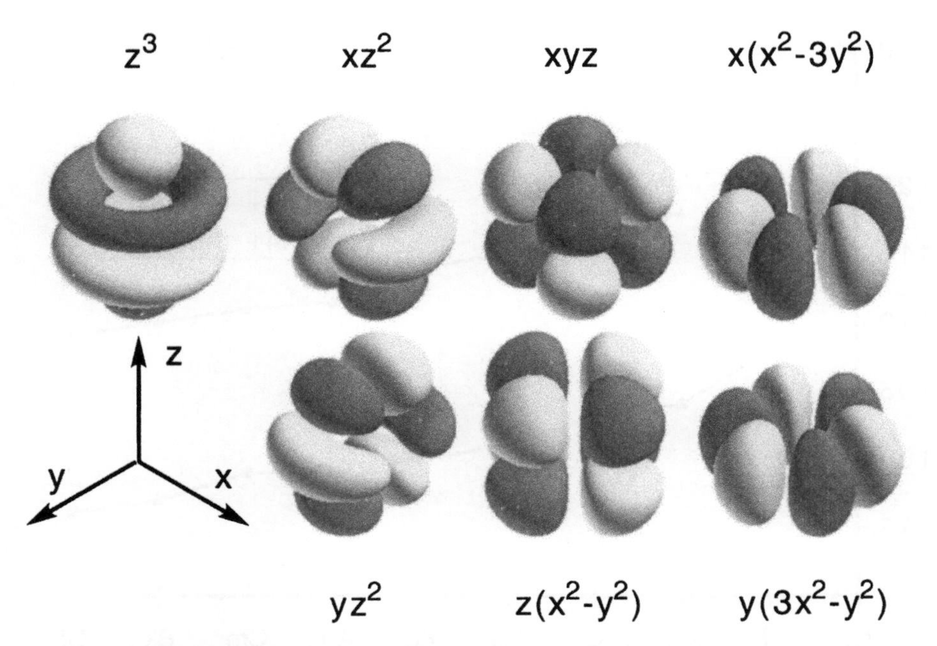

FIGURE 1.3 THE 'GENERAL' SET OF SEVEN F ORBITALS TOGETHER WITH THEIR SHORTENED CARTESIAN LABELS (D.L. COOPER, UNIVERSITY OF LIVERPOOL). THE 'CUBIC' SET CONSISTS OF $f(z^3)$ AND THE SIMILAR $f(x^3)$ AND $f(y^3)$; $f(z(x^2-y^2))$ AND THE SIMILAR $f(x(z^2-y^2))$ AND $y(z^2-x^2)$; AND $f(xyz)$.

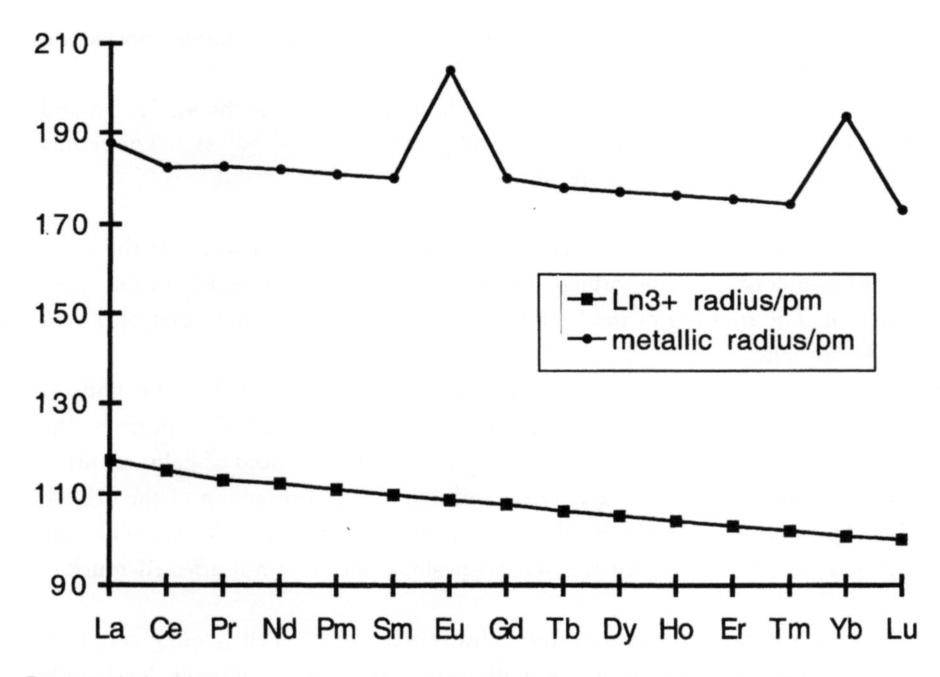

FIGURE 1.4 METALLIC AND IONIC RADII FOR THE LANTHANIDE SERIES.

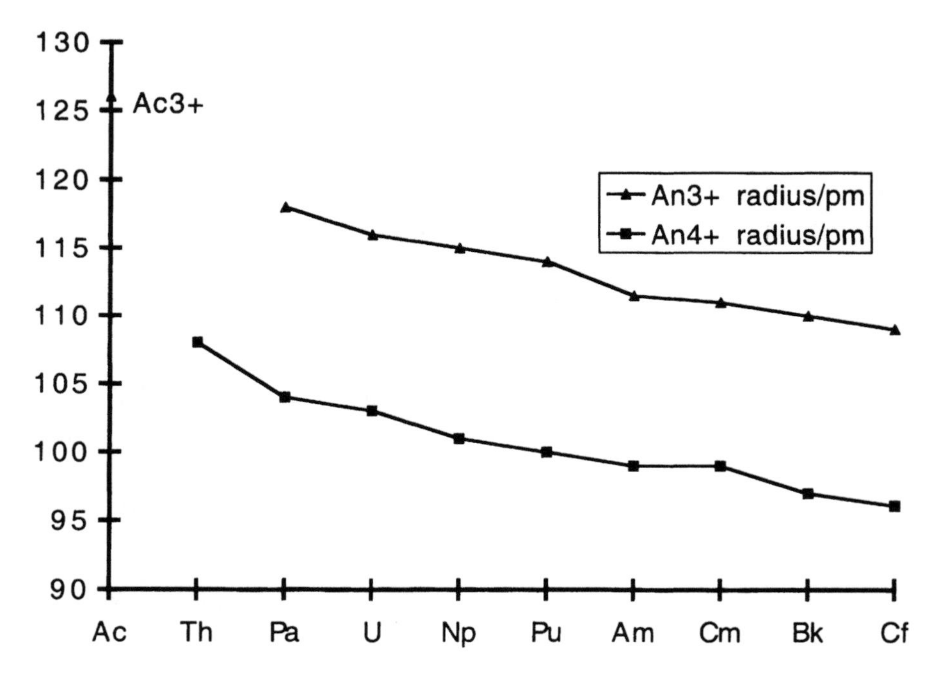

FIGURE 1.5 IONIC RADII FOR THE ACTINIDE SERIES.

the lanthanide contraction and between 40% and 50% of the actinide contraction is due to relativistic effects.

Figure 1.6 shows plots of the radial probability functions for the 4f, 5p, 5d and 6s orbitals of Lu which has an electron configuration [Xe]$6s^2$ $5d^1$ $4f^{14}$. The results of including relativistic effects are to:

- split the p, d and f orbitals by spin-orbit coupling in such a way that the inner of the two corresponding orbitals (smallest radial extension) is always the lower j orbital. The splitting of the 5p orbitals is more significant than that of the 5d or 4f orbitals.
- contract the s orbitals. This is because the s electrons, which have no node at the nucleus must have very high speeds in order to prevent their capture by the nucleus. For heavy atoms, this speed may be close to the speed of light, resulting in an increase in mass of the electron and hence to a contraction of the orbital. The $5p_{3/2}$ orbital is not very different from the nonrelativistic 5p, while the $5p_{1/2}$ is contracted. 5d expands relativistically while 4f is not affected much.

The plots in Figure 1.6 demonstrate clearly that the 4f shell is quite core-like. As a result the 4f-orbitals play essentially no part in the bonding in lanthanide complexes, and crystal field effects, which are so important for the d-transition metals, are negligible for the lanthanides. The 5f orbitals of the actinides have greater radial extension than the 4f orbitals of the lanthanides and are more effectively shielded from the nucleus. There is some evidence for 5f orbital participation in bonding for complexes of the early actinides.

1.4.4 Electronic Configurations

Lanthanides

The electron configurations, determined by electronic spectroscopy, of the gas phase lanthanide atoms and ions, are summarized in Table 1.6. The filling of the 4f orbitals in the atoms is fairly regular; the $4f^75d^16s^2$ configuration of atomic Gd is a result of the stability of the half-filled 4f shell. In the Ln^{3+} ions, the 4f orbitals are filled sequentially from La^{3+} ($4f^0$) to Lu^{3+} ($4f^{14}$).

The interaction of the f-orbitals of a lanthanide complex with the crystal field is very small and so in discussing electronic energies of these complexes it is reasonable to deal with free ion energies and then apply the crystal field at the end. All lanthanide complexes are weak-field high-spin complexes. The electronic states of lanthanide ions can be described using the Russell-Saunders (L-S) coupling scheme and the ground state term symbols can be derived using Hund's Rules:

- The term with the highest multiplicity ($2S_{max}+1$) has the lowest energy.
- If there are several terms with the same multiplicity, the term with the highest L has the lowest energy.
- If the sub-shell is less than half full then the level with the lowest J-value has the lowest energy; if the sub-shell is more than half full then the level with the highest J-value has the lowest energy.

For example, Pr^{3+} has a $4f^2$ configuration. The two electrons will occupy different orbitals in order to maximise S, resulting in a multiplicity of 3. The maximum value of L is achieved by placing the electrons in the orbitals with angular momentum 3 and 2 to give a total L = 5. The term of the ground state is derived from the relation:

L	**0**	**1**	**2**	**3**	**4**	**5**	**6**	**7**	**8**
Term symbol	S	P	D	F	G	H	I	J	K

and the ground state term symbol is therefore ^3H. The spin and orbital angular momenta can then interact by spin-orbit coupling to give a resultant J. J can take the values $|L + S|$, $|L + S - 1|$ through to $|L - S|$ and so for Pr^{3+} the ^3H term is split into three levels with J = 6, 5, 4. Because Pr^{3+} has a less than half filled sub-shell, the level with minimum J has the lowest energy and so the ground state term symbol for Pr^{3+} is $^3H_4$.

There is very little interaction between f-orbitals and the crystal field in a lanthanide complex and so there is very little difference between energy levels in a complex and in the free ion.

Actinides

The electron configurations, determined by electronic spectroscopy, of the gas phase actinide atoms are summarized in Table 1.7. The filling of the 5f orbitals in the actinide

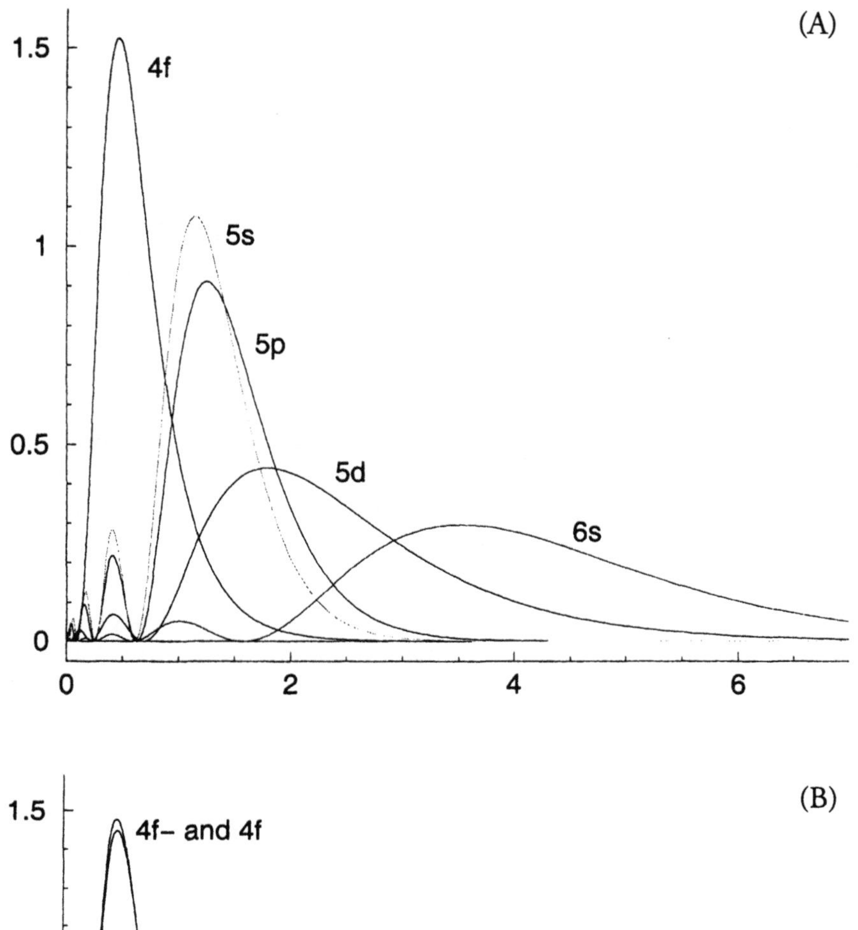

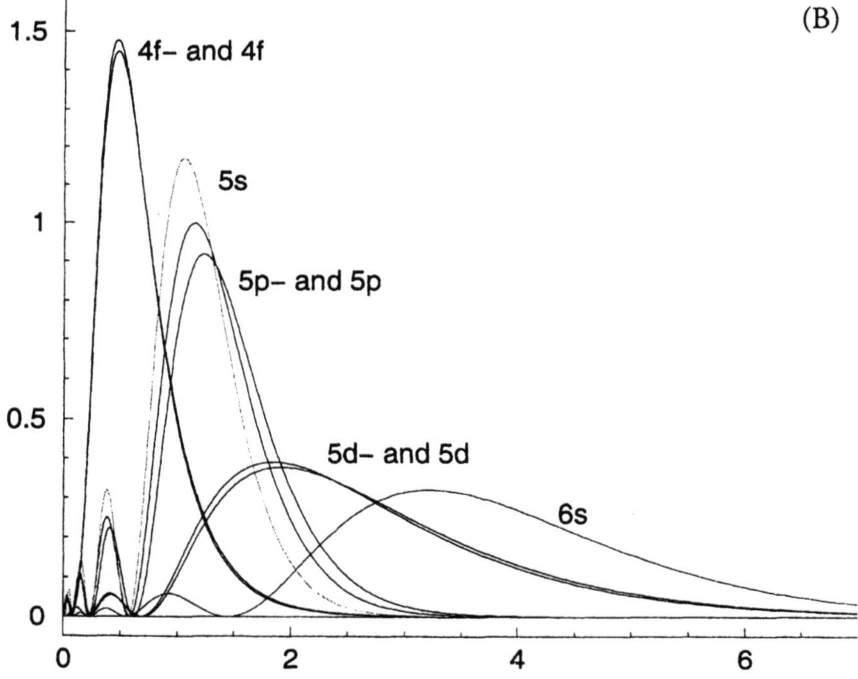

FIGURE 1.6 RADIAL PROBABILITY FUNCTIONS FOR THE 4F, 5P, 5D AND 6S ORBITALS OF LU (A) EXCLUDING RELATIVISTIC EFFECTS (B) INCLUDING RELATIVISTIC EFFECTS (JON K. LAERDAHL, UNIVERSITY OF AUCKLAND).

series is not quite as regular as the filling of the lanthanide 4f orbitals. In Th the 6d orbitals are lower in energy than the 5f and so Th, unlike its lanthanide analogue Ce, has no f electron. After Th, the 5f shell becomes more stable than the 6d shell. For the elements Ac to Np the $5f^{N-1}ds^2$ configuration is more stable than the f^Ns^2 configuration; the actinides beyond Np have similar electron configurations to those of their 4f analogues. The 5f electrons are shielded from the nucleus to a greater extent than the 4f electrons of the lanthanide series and so the energy differences between 5f, 6d and 7s are much smaller than the corresponding differences for the lanthanides.

For actinide complexes, spin-orbit coupling is considerably greater, and the greater radial extension of the 5f orbitals, especially for the early actinides, results in larger crystal field splittings and there is often considerable overlap between levels of different states.

1.4.5 Magnetic properties

Because there is significant spin-orbit coupling in ions of lanthanides and actinides, the 'spin-only' formula which works well for many first row transition metal complexes is not applicable to f-element complexes. The effective magnetic moment for these elements depends on J, where $J = L + S$.

$$\mu_{\text{eff}} \propto \sqrt{J(J+1)}$$

The constant of proportionality is the Landé g value, g where g is given by

$$g = 1 + \frac{S(S+1) + J(J+1) - L(L+1)}{2J(J+1)}$$

This model works well for complexes of the lanthanides as shown by the data in Table 1.8, the only exceptions being Eu^{3+} and Sm^{3+}, both of which have low-lying excited states which are appreciably populated at room temperature. The magnetic moments of both these ions decrease with decreasing temperature and Eu^{3+} complexes are diamagnetic at low temperatures. When population of excited states is included in the calculation of μ_{eff}, good agreement with experiment is achieved. The magnetic properties of actinides are not so simple because of increased crystal field effects. A detailed account of magnetic properties of actinide compounds has been given by Edelstein and Goffart (1986).

1.4.6 Oxidation States

Lanthanides

In general the +3 oxidation state is the most stable for all the lanthanides. In fact, according to Pimentel (1971) 'Lanthanum has only one important oxidation state in aqueous solution, the +3 state. With few exceptions, that staement tells the whole boring story about the other fourteen elements.' However, this is by no means the whole story and oxidation states +2 or +4 are known for several lanthanides, especially in the solid state.

TABLE 1.8 MAGNETIC PROPERTIES OF Ln^{3+} IONS.

Ln	4f configuration	ground state term symbol	g	$g_J\sqrt{(J(J+1))}$	experimental μ_{eff} for $[Ln(NO_3)_3(phen)_2]$*
La^{3+}	$4f^0$	$^1S_0$	0	0	0
Ce^{3+}	$4f^1$	$^2F_{5/2}$	6/7	2.54	2.46
Pr^{3+}	$4f^2$	$^3H_4$	4/5	3.58	3.48
Nd^{3+}	$4f^3$	$^4I_{9/2}$	8/11	3.62	3.44
Pm^{3+}	$4f^4$	$^5I_4$	3/5	2.68	
Sm^{3+}	$4f^5$	$^6H_{5/2}$	2/7	0.84	1.64
Eu^{3+}	$4f^6$	$^7F_0$	1	0.0	3.36
Gd^{3+}	$4f^7$	$^8S_{7/2}$	2	7.94	7.97
Tb^{3+}	$4f^8$	$^7F_6$	3/2	9.72	9.81
Dy^{3+}	$4f^9$	$^6H_{15/2}$	4/3	10.63	10.6
Ho^{3+}	$4f^{10}$	$^5I_8$	5/4	10.60	10.7
Er^{3+}	$4f^{11}$	$^4I_{15/2}$	6/5	9.59	9.46
Tm^{3+}	$4f^{12}$	$^3H_6$	7/6	7.57	7.51
Yb^{3+}	$4f^{13}$	$^2F_{7/2}$	8/7	4.54	4.47

*Values from Hart, F.A. and Laming, F.P. (1965) *J. Inorg. Nucl. Chem.*, 27, 1605–1610.

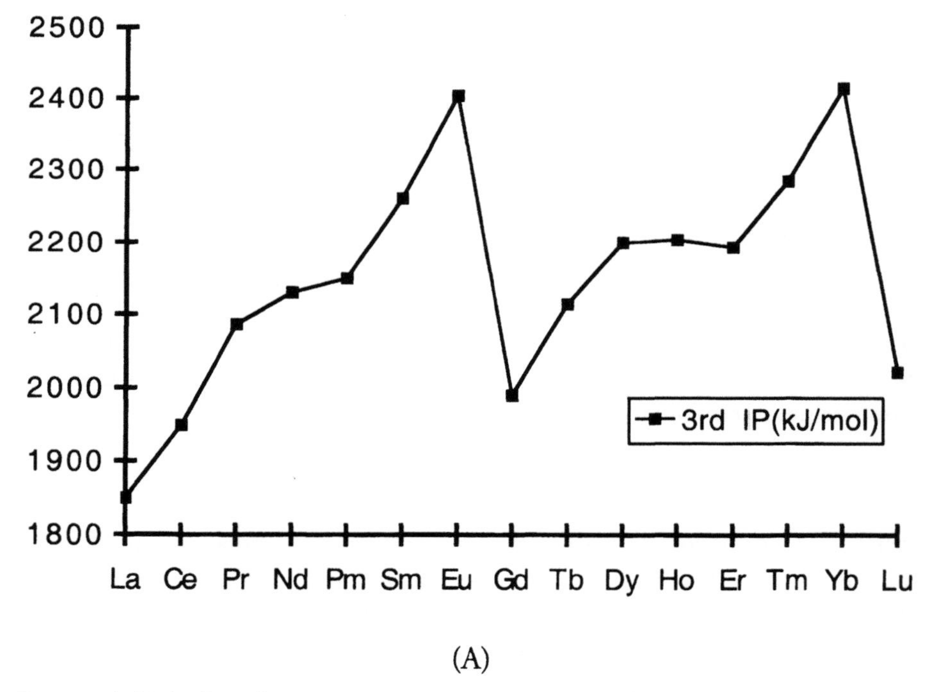

(A)

FIGURE 1.7(A) THE 3RD IONIZATION POTENTIAL FOR LN.

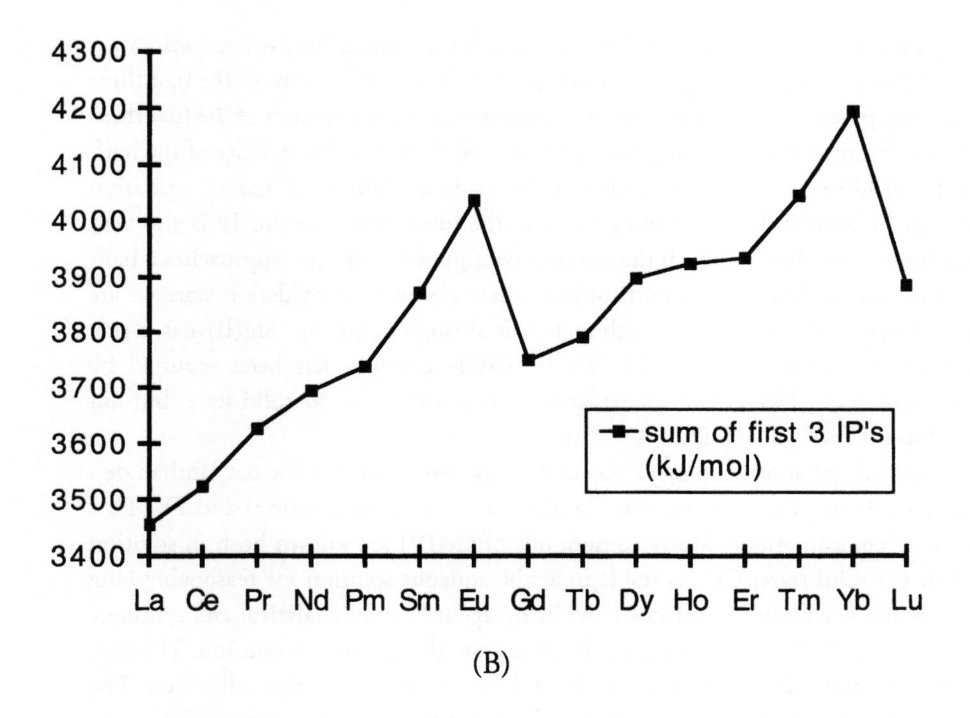

(B)

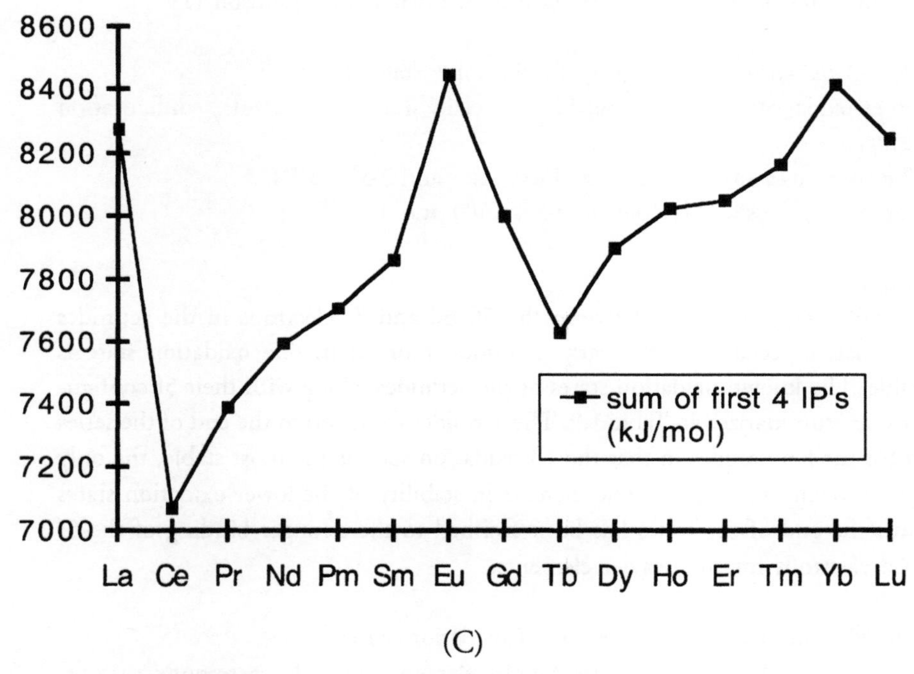

(C)

FIGURE 1.7(B,C) IONIZATION POTENTIALS FOR LN (B) THE SUM OF THE FIRST THREE IONIZATION POTENTIALS (C) THE SUM OF THE FIRST FOUR IONIZATION POTENTIALS.

Figure 1.7 (a) shows a plot of the 3rd ionization potential for the lanthanides *i.e.* for the process $Ln^{2+}_{(g)} \rightarrow Ln^{3+}_{(g)} + e^-$ and Figure 1.7 (b) shows the sum of the first three ionization potentials. The 3rd ionization potential, and also the sum of the first three IP's, are at a maximum for Eu^{2+} ($4f^7$) and Yb^{2+} ($4f^{14}$) due to the stability of the half-filled and filled 4f shell. This results in the ready availability of the +2 oxidation state for Eu and Yb both in solution and in the solid state. The 3rd IP is also very high for Sm and Tm for which the electron configuration of Ln^{2+} approaches a half-filled or full 4f shell. Compounds of both these elements in oxidation state +2 are well known in the solid state. Although it is strongly reducing, Sm(II) has a rich solution chemistry, and a soluble Tm(II) iodide complex has been reported by Bochkarev *et al.* (1997). The +2 oxidation state is known in the solid state (but not in solution) for Nd and Dy.

Figure 1.7 (c) shows a plot of the sum of the first four IP's for the lanthanides. This plot shows clear minimum at Ce (Ce^{4+} has a $4f^0$ configuration) and Tb (Tb^{4+} has a $4f^7$ configuration). Many compounds of Ce(IV) are known both in solution and in the solid state. Ce^{4+} is stable in acidic aqueous solution for reasonably long periods and is a useful one-electron oxidizing agent both in analytical and synthetic chemistry. Tb(IV) is well known in the solid state though not in solution. The sum of the first four IP's is also low for Pr and Pr^{4+} is known in the solid state. The oxidation states of the lanthanides have been discussed by Johnson (1977).

- The +3 oxidation state is generally the most stable for Ln.
- Other oxidation states are available, especially if a $4f^0$, $4f^7$ or $4f^{14}$ configuration results.
- The most accessible 2+ ions are Eu^{2+} ($4f^7$) and Yb^{2+} ($4f^{14}$).
- The most accessible 4+ ions are Ce^{4+} ($4f^0$) and Tb^{4+} ($4f^7$).

Actinides

The small energy difference between the 5f, 6d and 7s electrons in the actinides means that, especially for the early actinides, more than one oxidation state is available. The known oxidation states of the actinides, along with their 5f configurations, are summarized in Table 1.9. The actinides from Am to the end of the series resemble the lanthanides in that the +3 oxidation state is the most stable, the only exception being No^{2+} ($5f^{14}$). The increase in stability of the lower oxidation states towards the end of the series has been ascribed to the stronger binding of the 5f and 6d electrons for the heavier elements.

- The elements Pa to Cm have several oxidation states.
- The most stable oxidation states for the elements Ac to U correspond to a $5f^0$ configuration.
- The +3 oxidation state is the most stable for Ac, and Am to Lr (except No).
- The most stable oxidation state for No is +2 (No^{2+} has $5f^{14}$ configuration).

TABLE 1.9 OXIDATION STATES AND ELECTRON CONFIGURATIONS FOR THE ACTINIDES.

	Ac	Th	Pa	U	Np	Pu	Am	Cm	Bk	Cf	Es	Fm	Md	No	Lr
+2														$5f^{14}$	
+3	$5f^0$				$5f^4$		$5f^6$	$5f^7$	$5f^8$	$5f^9$	$5f^{10}$	$5f^{11}$	$5f^{12}$		$5f^{14}$
+4		$5f^0$		$5f^2$		$5f^4$					?				
+5			$5f^0$		$5f^2$			?							
+6				$5f^0$				?							
+7							?								

Common oxidation state

Known oxidation state

? Unconfirmed

1.5 BINARY COMPOUNDS

This section will survey the oxides and halides of the f-block elements, illustrating the range of available oxidation states, and the structural trends. Because the bonding is essentially ionic in these compounds, their structures are determined almost entirely by steric factors and structural changes across the series can be correlated with changes in ionic radii. Lanthanide halides have been reviewed by Burgess and Kijowski (1981). Structural features of actinide halides (and especially those of U) have been reviewed by Taylor (1976).

1.5.1 Oxides of the lanthanides

The lanthanide sesquioxides Ln_2O_3 are amongst the most thermodynamically stable oxides known ($\Delta_f G°$ for La_2O_3 = –1705.8 kJmol^{-1} *c.f.* –944 kJmol^{-1} for TiO_2 and –1582.3 kJmol^{-1} for Al_2O_3). They are the stable oxides for all the Ln elements except Ce, Pr and Tb, and are the final product of calcination of many lanthanide compounds such as oxalates, nitrates and carbonates. Their great stability is a consequence of the stability of the Ln(III) oxidation state and the oxophilicity of the Ln^{3+} ions. The stable oxide of cerium is CeO_2 (ceria), which is formed on calcination in air of most Ce salts (Ce(III) or Ce(IV)). Like Ce, Pr and Tb also have accessible +4 oxidation states and their most stable oxides are the mixed valence species Pr_6O_{11} and Tb_4O_7, both of which are very intensely coloured due to charge transfer transitions. Sesquioxides of Ce, Pr and Tb are stable only under strongly reducing conditions.

Structures of oxides are often understood in terms of M^{n+} ions occupying holes in close-packed arrays of O^{2-} ions, but because of the large size of Ln^{3+} ions (7-coordinate radius of La^{3+} = 124 pm) compared with O^{2-} (4-coordinate radius

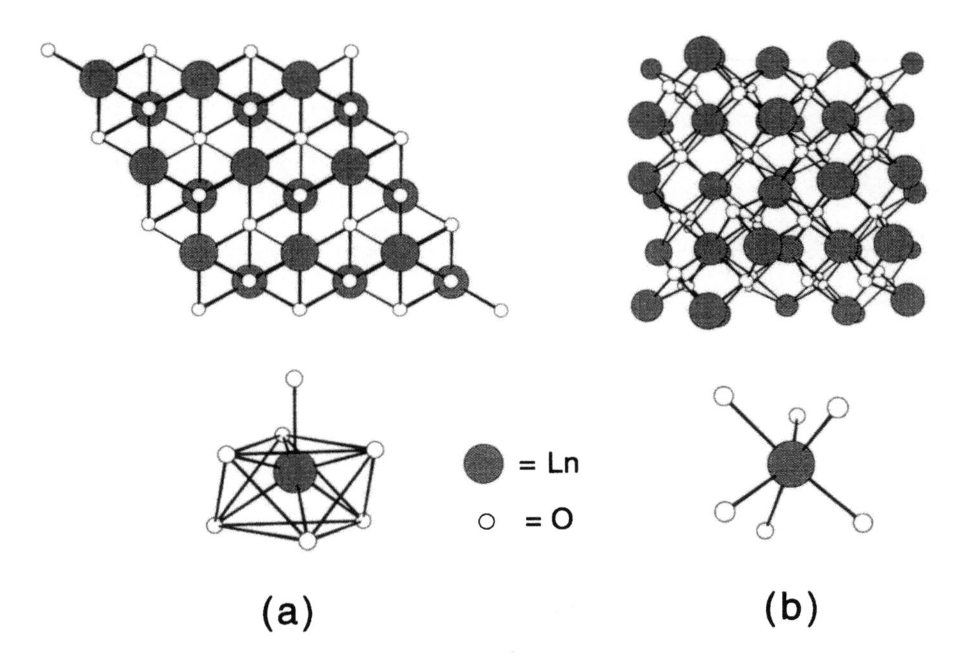

FIGURE 1.8 (A) Ln_2O_3 A-FORM (B) Ln_2O_3 C-FORM.

= 124 pm) this simple picture is not appropriate for the lanthanides. Ln_2O_3 adopt one of three structures in the solid state. The type A (hexagonal) structure is adopted for the large, early lanthanides and contains 7-coordinate Ln atoms which are surrounded by a mono-capped octahedral array of O^{2-} ions. The type B (monoclinic) structure is related to type A, but is more complex and contains three types of non-equivalent Ln atoms, some with octahedral 6-coordination and some with monocapped trigonal prismatic 7-coordination. The later, heavier, Ln adopt the type-C (cubic) structure which is related to the fluorite structure with removal of 1/4 of the anions in a regular manner. CeO_2 adopts the fluorite structure, and the structures of Pr_6O_{11} and Tb_4O_7 are closely related to fluorite.

Lanthanide oxides absorb atmospheric moisture and CO_2; this reactivity is much more facile for type A or B oxides than for type C oxides, a trend which is also observed in the reaction of Ln_2O_3 with aqueous acids.

1.5.2 Halides of the lanthanides

Fluorides

Trifluorides are known for all the lanthanides. They are readily prepared by the action of HF on Ln_2O_3 or $Ln_2(CO_3)_3$; if the preparation is carried out under aqueous conditions the product is partially hydrated $LnF_3 \cdot 1/2H_2O$, which is virtually insoluble in H_2O.

Two types of structure are adopted by LnF_3, depending on the ionic radius of Ln^{3+}. The early LnF_3 adopt the tysonite structure (after the mineral tysonite which is a mixed LnF_3). The description of the structure is not straightforward: in LaF_3

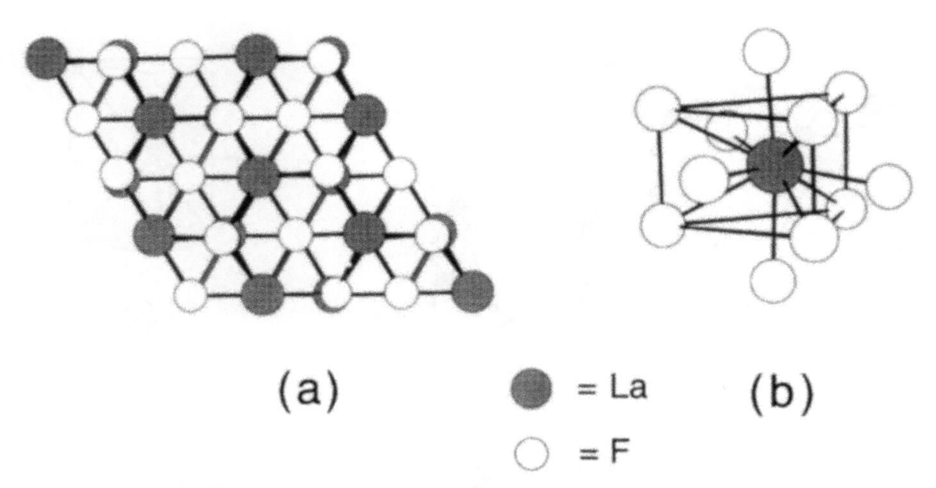

FIGURE 1.9 (A) STRUCTURE OF LaF_3 (B) THE LaF_{11} FRAGMENT.

the La is 11-coordinate with a fully-capped distorted trigonal prismatic coordination geometry. The later LnF_3 (Sm -Lu) have 9-coordinate Ln surrounded by a tri-capped trigonal prism of F atoms.

Lanthanide trifluorides are of technological importance: they are the preferred precursor for production of Ln metal by reduction with Ca as they can be prepared anhydrous and oxygen free. They are also important as coatings for optical lenses, and the tysonite fluorides have high fluoride mobility leading to ionic conductivity.

Tetrafluorides LnF_4 are known for Ce, Pr and Tb, the elements for which the +4 oxidation state is most accessible. They are prepared by fluorination with F_2 of the metal (in the case of Ce) or the trifluoride. Unstable LnF_2 are known for Sm, Eu and Yb, which have the most stable +2 oxidation states.

Chlorides

Lanthanide trichlorides $LnCl_3$ are important starting materials in lanthanide coordination chemistry and can also be used as starting materials for preparation of the metals either by electrolytic reduction or chemical reduction with Ca. Hydrated $LnCl_3$ are easily prepared by the action of aqueous HCl on the appropriate oxide (CeO_2, Pr_6O_{11} and TbO_7 all oxidise HCl to Cl_2 and yield $LnCl_3$). Preparation of the anhydrous $LnCl_3$ (which are extremely hygroscopic) is not nearly so straightforward: simply heating the hydrated $LnCl_3$ results in elimination of HCl and formation of the oxychloride LnOCl. The most widely used laboratory route to anhydrous $LnCl_3$ is to heat up the $LnCl_3$ *in vacuo* in the presence of NH_4Cl (which suppresses elimination of HCl) and then to remove NH_4Cl by sublimation.

The early $LnCl_3$ adopt the 'UCl_3' structure shown in Figure 1.10 in which the large Ln^{3+} ion is coordinated to nine approximately equidistant Cl neighbours in a tri-capped trigonal prismatic arrangement. Later $LnCl_3$ adopt the 'YCl_3' layer structure in which the smaller Ln^{3+} ions are surrounded by an octahedron of Cl neighbours.

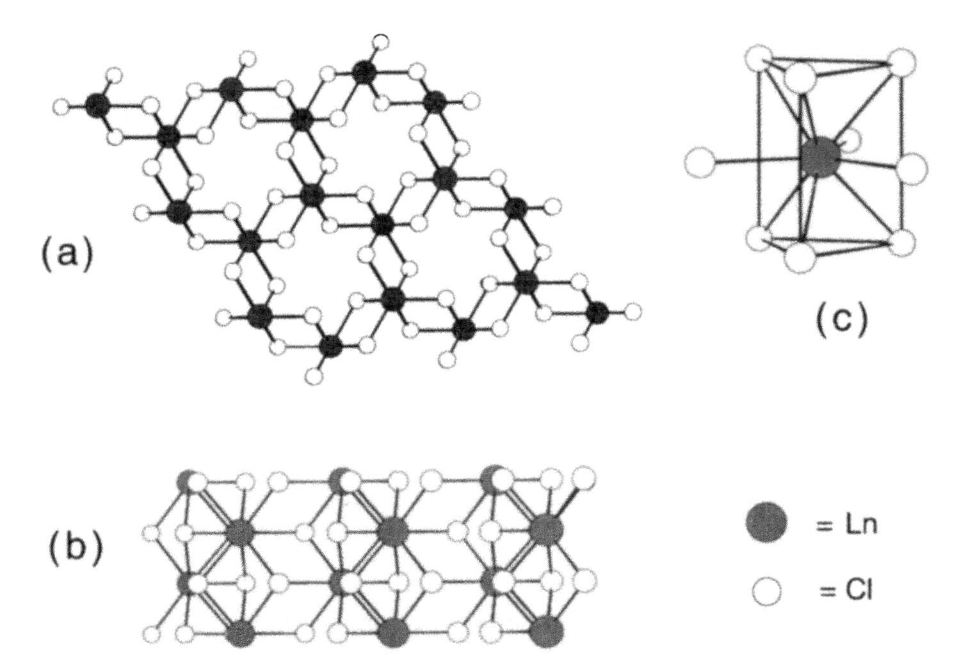

FIGURE 1.10 THE 'UCL3' STRUCTURE ADOPTED BY EARLY LNCL3 (a) VIEW ALONG c; (b) VIEW ALONG *a*, AND (c) $LnCl_9$ FRAGMENT.

Chlorides are also known in oxidation states lower than +3: ionic dichlorides are known for Nd, Sm, Eu, Dy and Tm, and a sesquichloride Ln_2Cl_3 was first reported for Gd by Lokken and Corbett (1973). This stucture is best formulated as $[Gd_4]^{6+}$ $[Cl^-]_6$, and is made up of infinite chains of edge-linked Gd_6 octahedra aligned parallel to the b axis. Each Gd is coordinated to four other Gd atoms (the nearest at 3.5 Å, the others at between 3.7 and 3.8 Å) and to five Cl atoms at between 2.72 and 2.88 Å. Three views of the Gd_2Cl_3 structure are shown in Figure 1.11.

Bromides and Iodides

The tribromides and triiodides are known for all the lanthanide elements. They are readily prepared as hydrated salts by reaction of the lanthanide oxide with the appropriate hydrohalic acid. The early tribromides (La to Pr) adopt the UCl_3 structure while the $PuBr_3$ structure is adopted by the later tribromides (Nd to Lu) and the early triiodides (La to Nd). As with the trichlorides, anhydrous LnI_3 and $LnBr_3$ cannot be obtained by simply heating the hydrated salts: the dehydration must be carried out in the presence of excess hydrogen halide or ammonium halide to prevent decomposition to the oxyhalide.

Ionic dibromides and diiodides are known for Nd, Sm, Eu, Dy, Tm and Yb. The diiodides of Sm, Eu and Yb are useful starting materials for organometallic and coordination chemistry of these elements in their +2 oxidation states, and SmI_2 has in recent years become a popular 1-electron reducing agent for organic chemistry (See Chapter 5). The diiodides of La, Ce, Pr and Gd have metallic properties and are best formulated as $Ln^{3+}(I^-)_2(e)$.

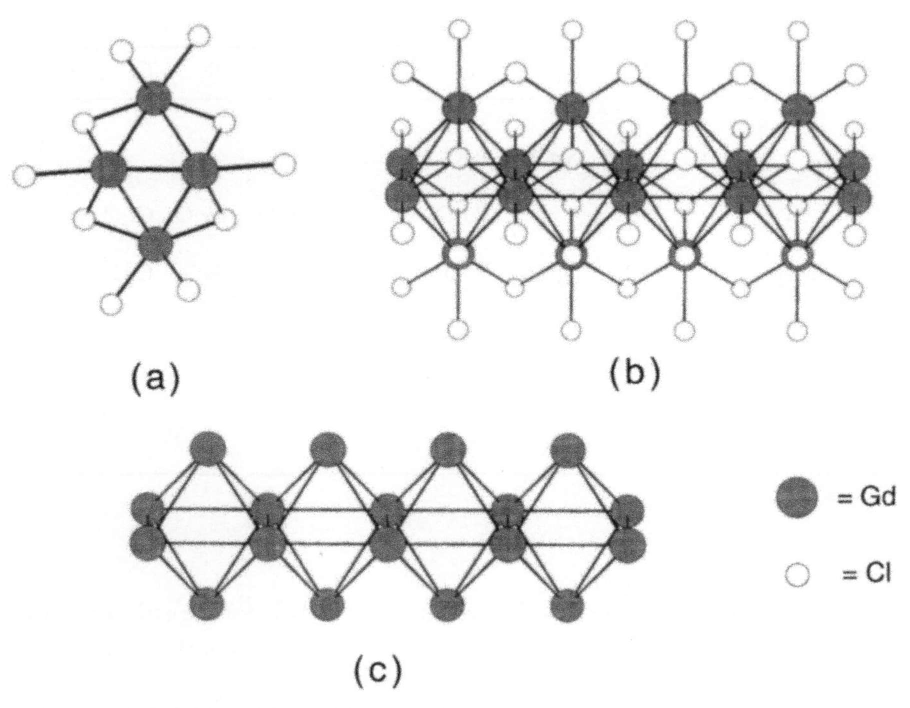

FIGURE 1.11 THE STRUCTURE OF Gd_2Cl_3 (A) VIEW OF A CHAIN ALIGNED ALONG THE B AXIS (B) VIEW PERPENDICULAR TO THE B AXIS (C) CHAIN OF EDGE-LINKED Gd_6 OCTAHEDRA.

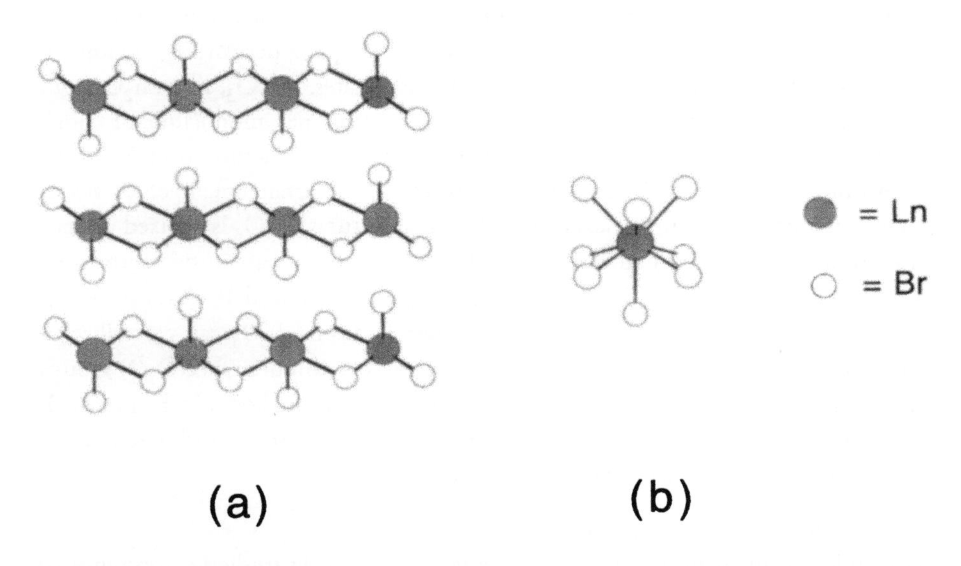

FIGURE 1.12 THE '$PuBr_3$' STRUCTURE ADOPTED BY LATER $LnBr_3$ AND EARLY LnI_3 (A) VIEW PERPENDICULAR TO THE LAYERS (B) THE $PuBr_8$ FRAGMENT.

TABLE 1.10 ACTINIDE OXIDES.

Ac	Th	Pa	U	Np	Pu	Am	Cm	Bk	Cf	Es
		PaO		NpO	PuO	AmO				
Ac_2O_3					Pu_2O_3	Am_2O_3	Cm_2O_3	Bk_2O_3	Cf_2O_3	Es_2O_3
	ThO_2	PaO_2	UO_2	NpO_2	PuO_2	AmO_2	AmO_2	BkO_2	CfO_2	
			U_4O_9	Np_2O_5						
			U_3O_8							
			UO_3	NpO_3						

1.5.3 Oxides of the actinides

Because of the range of available oxidation states, particularly for the early actinides, the chemistry of actinide oxides is complex. The known oxides of the actinides are summarized in Table 1.10.

Probably the most important of these is the dioxide AnO_2, which is known for An = Th to Cf. AnO_2, like LnO_2, crystallize with the fluorite structure. This structure is rather open, and as a result extra O atoms can be absorbed to give AnO_{2+x} without changing the overall structure. For U, additional O can be absorbed up to x=0.25, resulting in U_4O_9, where the extra O atoms are ordered to give a superlattice with a unit cell 64 times the volume of the UO_2 cell. The sesquioxides An_2O_3, are known for Ac and Pu – Es, which have stable +3 oxidation states. An_2O_3, like Ln_2O_3, can crystallize in one or more of three structural types: hexagonal, monoclinic or cubic. The coordination number of An^{3+} is 7 for the hexagonal structure, 6 or 7 for the monoclinic structure and 6 for the cubic structure; the arrangement of O atoms around An is not very symmetrical in any of these structures. U_3O_8 is a mixed valence compound with two U(V) atoms and one U(VI) atom. It occurs in two forms: the more common α form has 7-coordinate (pentagonal bipyramidal) U atoms; the β form has two 7-coordinate and one 6-coordinate U atoms. UO_3 can crystallize in six forms. The high-pressure form has been shown to contain uranyl (UO_2^{2+}) groups with 5 other O atoms coordinated in the equatorial plane as shown in Figure 1.13. The actinide monoxides adopt the NaCl structure, however it is possible that some of these species may actually be oxycarbides.

1.5.4 Halides of the actinides

The halides of the actinides are among the most thoroughly studied compounds of these elements. As with the oxides, a wide range of compounds is known, and these are summarized in Table 1.11. Structural features of actinide halides (and especially those of U) have been reviewed by Taylor (1976).

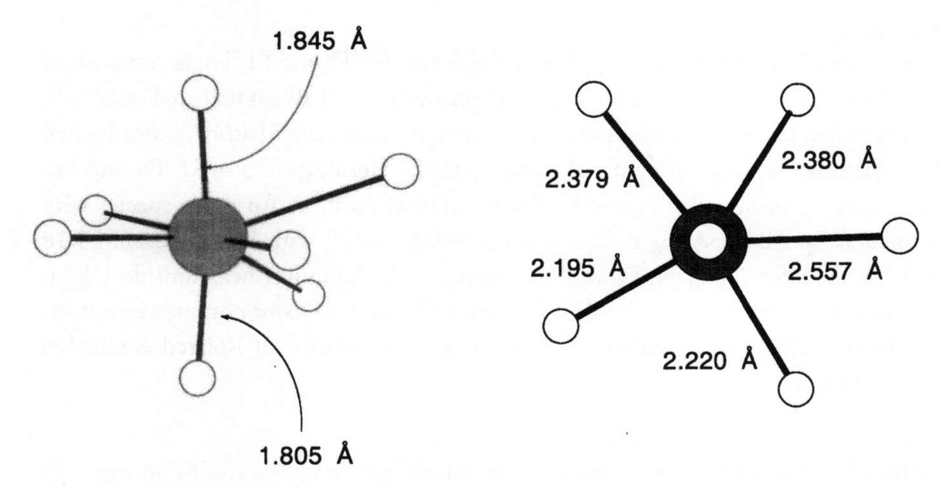

FIGURE 1.13 TWO VIEWS OF THE UO_7 FRAGMENT IN THE HIGH-PRESSURE FORM OF UO_3.

TABLE 1.11 STRUCTURALLY CHARACTERIZED ACTINIDE HALIDES.

	Ac	Th	Pa	U	Np	Pu	Am	Cm	Bk	Cf	Es
AnX_2							Cl				
							Br			Br	
	I						I			I	
AnX_3	F			F	F	F	F	F	F	F	
	Cl			Cl	Cl	Cl	Cl	Cl	Cl	Cl	Cl
	Br			Br	Br	Br	Br	Br	Br	Br	Br
				I	I	I	I	I	I	I	
AnX_4		F	F	F	F	F					
		Cl	Cl	Cl	Cl						
		Br	Br	Br	Br						
		I		I							
An_2X_9			F	F							
AnX_5			F	F	F						
			Cl	Cl							
			Br	Br							
AnX_6				F	F	F					
				Cl							

Fluorides

Trifluorides AnF_3 are known for all the actinides except Th and Pa. These compounds all adopt the LaF_3 structure as described previously and illustrated in Figure 1.9. The tetrahalides AnF_4 all adopt the UF_4 structure where each U atom is coordinated by a distorted square antiprism of eight F atoms. Hexafluorides of U, Pu and Np are known; they are all prepared by fluorination of AnF_4 or An oxide species with elemental F_2. All the AnF_6 are low melting volatile solids with bp's of 62.26, 55.18 and 56.54°C for PuF_6, NpF_6 and UF_6 respectively. Uranium hexafluoride UF_6 is the most thoroughly studied halide of uranium because of its importance in isotope separation. The crystal structure of UF_6 shows it to consist of isolated octahedral UF_6 molecules.

UF_6 in uranium isotope separation

Naturally occurring uranium consists of two major isotopes: ^{235}U (0.711%) and ^{238}U (99.28%). Most nuclear reactors require the ^{235}U concentration to be enriched up to between 3 and 5%, and although there are several possible processes which can be used, in practice only two — gaseous diffusion and high speed gas centrifugation — are employed commercially. Both of these processes require a U compound which is gaseous under convenient conditions. UF_6 fulfils this criterion (its bp is 56.54°C) and has the further advantage that F has only one isotope (^{19}F).

Gaseous diffusion of UF_6 was the first process to be used for large scale uranium enrichment. It relies on the different rates of diffusion through a membrane of $^{235}UF_6$ (M_r = 349.03) and $^{238}UF_6$ (M_r = 352.04). Graham's law states that:

$$\frac{\text{rate of diffusion of } ^{235}UF_6}{\text{rate of diffusion of } ^{238}UF_6} = \sqrt{\frac{352.04}{349.03}} = 1.0043$$

The separation factor of 1.0043 means that in order to achieve satisfactory enrichment many separation stages must be linked in a cascade. Figure 1.14 shows the essential features of a stage in a diffusion cascade; a typical diffuser unit is a cylinder approximately 3.5 m in diameter and 6 m long.

The gas centrifuge process for isotope separation depends on the absolute mass difference between $^{235}UF_6$ and $^{238}UF_6$ rather than on the square root of the relative mass difference as in the diffusion technique, and is very sensitive to the peripheral velocity of the centrifuge. The process results in an enrichment in the heavier isotope towards the periphery of the centrifuge and an enrichment in the lighter isotope in the centre. The enriched $^{235}UF_6$ is removed by scooping from the centre of the centrifuge.

The laser separation of $^{235}UF_6$ and $^{238}UF_6$ has been known for many years but has not been commercially exploited. In this process a ν(U-F) vibration of $^{235}UF_6$ is selectively excited using an IR laser. The excited $^{235}UF_6$ molecule is then dissociated by a UV laser to give involatile $^{235}UF_5$ particles, which can easily be separated, and F_2 gas. The $^{238}UF_6$ molecules are not in a vibrationally excited state and are not decomposed by UV irradiation.

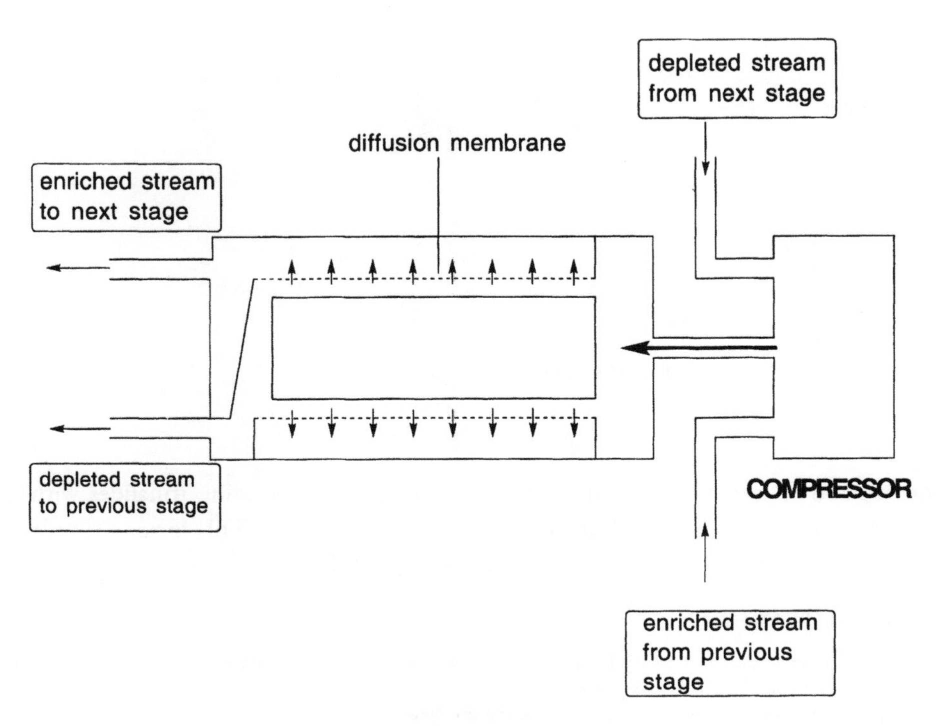

FIGURE 1.14 SCHEMATIC DIAGRAM OF A DIFFUSER UNIT USED FOR UF_6 SEPARATION.

Chlorides

Trichlorides $AnCl_3$ are known for all the actinides except Th and Pa, and can be prepared by the high temperature reaction of the oxides with HX. Dehydration of the hydrated trihalides is easier than for the Ln analogues. They all adopt the UCl_3 structure shown in Figure 1.10 in which each U atom is coordinated to nine Cl atoms in a tricapped trigonal prismatic arrangement. $AnCl_4$ are available for Th to Np, and are isostructural with the tetrafluorides. $AnCl_5$ are known for Pa and U; UCl_5 is much less stable than $PaCl_5$, disproportionating above 100°C to give UCl_4 and UCl_6. The structure of UCl_5 is based on molecular U_2Cl_{10} dimers; that of $PaCl_5$ consists of infinite chains of edge-sharing pentagonal bipyramids as shown in Figure 1.15. UCl_6 is the only actinide hexachloride ; it melts at 177.5°C, it is moderately volatile and the Cl-Cl distances in the solid are consistent with a molecular species. Its structure is based on a hcp array of Cl atoms with U in 1/6 of the octahedral holes.

Bromides and Iodides

Triiodides AnI_3 are known for Th to Cf; ThI_3 disproportionates to ThI_2 and ThI_4, and tribromides $AnBr_3$ are known for all the actinides except Th and Pa. The $PuBr_3$ structure is adopted for all the triiodides and for the tribromides of Np to Bk; it is closely related to the UCl_3 structure with one of the 'capping' halogen atoms removed from the An coordination sphere to give an 8-coordinate bi-capped trigonal prism. $AcBr_3$ and UBr_3 both adopt the UCl_3 structure. Dibromides and diiodides

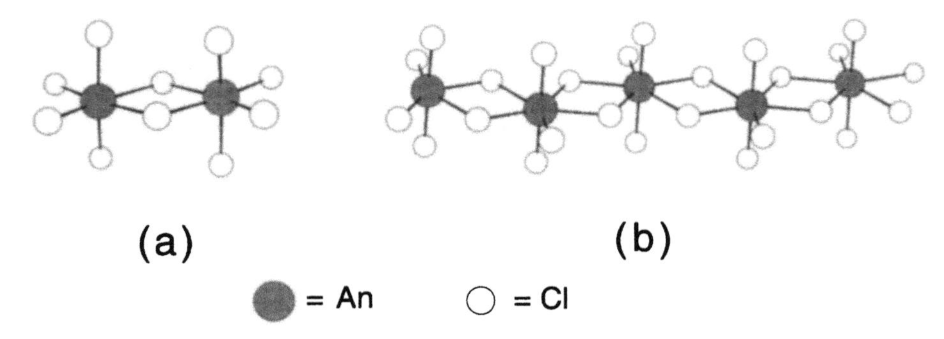

FIGURE 1.15 STRUCTURES OF $AnCl_5$ (A) UCl_5 (B) $PaCl_5$.

of Am and Cf can be obtained by reduction of the corresponding trihalides with H_2; they are isostructural with the corresponding Eu dihalides. ThI_2 is not a simple Th(II) dihalide: it is better considered as $[Th^{4+}](e^-)_2(I^-)_2$.

GENERAL READING

Two concise texts on the chemistry of the lanthanides and actinides have been published:
Cotton, S.A. *Lanthanides and Actinides*, Macmillan, London, 1991.
Kaltsoyannis, N., Scott, P. *The f elements*. Oxford University Press, 1999.
An interesting account of the history of the lanthanides can be found in:
Evans, C.H. *Episodes from the history of the rare earth elements*. Dordrecht London Kluwer 1996.

Very comprehensive surveys are available in:
Gschneider, K.A. Jr.; Eyring, L. *Handbook on the physics and chemistry of rare earths*. North Holland, Amsterdam 1978 –1987.
Freeman, A.J., Keller, C. *Handbook on the physics and chemistry of the actinides*. North Holland, Amsterdam, 1985.
Katz, J.J., Seaborg, G.T., Morss, L.R. *The chemistry of the actinide elements*. Chapman and Hall, London, 1986.

Comprehensive accounts of coordination chemistry of the lanthanides and actinides have been written:
Hart, F.A. (1987) 'The Lanthanides' and Bagnall, K.W. (1987) 'The Actinides' In *Comprehensive Coordination Chemistry*, Volume 3. Edited by G. Wilkinson, R.D. Gillard, J.A. McCleverty. Pergamon.

A comprehensive review of the organometallic chemistry of the lanthanides and actinides is:
Edelman, F. T. (1995) Scandium, yttrium, and the lanthanide and actinide elements, excluding their zero oxidation states. In *Comprehensive Organometallic Chemistry*, Volume 4, Edited by E.W. Abel, F.G.A. Stone and G. Wilkinson. Elsevier.

REFERENCES

Bochkarev, M.N., Fedushkin, I.L., Fagin, A.A., Petrovskaya, T.V., Ziller, J. W., Broomhall-Dillard, R.N.R. and Evans, W.J. (1997) Synthesis and structure of the first molecular thulium(II) complex: $[TmI_2(MeOCH_2CH_2OMe)_3]$. *Angew. Chem., Int. Ed.*, **36**, 133–135.
Burgess, J. and Kijowski, J. (1981) Lanthanide, yttrium and scandium trihalides: preparation of anhydrous materials and solution thermochemistry. *Adv. Inorg. Radiochem.*, **24**, 57–114.
Carter, C.P. and Taylor, M.D. (1962) Preparation of anhydrous lanthanide halides, especially iodides *J. Inorg. Nucl. Chem.*, **24**, 387.
Cotton, S.A. (1999) Element 61 – the missing lanthanide. *Education in Chemistry*, 96–98.
Johnson, D.A. (1977) Recent advances in the chemistry of the less-common oxidation states of the lanthanide elements. *Adv. Inorg. Chem. Radiochem.*, **20**, 1–132.
Laerdahl, J.K., Faegri, Jr., J., Visscher, L. and Saue, T. (1998) A fully relativistic Dirac-Hartree-Fock and second order Møller-Plesset study of the lanthanide and actinide contraction. *J. Chem. Phys.*, **109**, 10806–10817.

Lokken, D.A. and Corbett, J.D. (1973) Rare earth metal halide systems. XV. Crystal structure of gadolinium sesquichloride. A phase with unique metal chains. *Inorg. Chem.*, 12, 556–559.
Pimentel, G.C. and Spratley, R.D. (1971) *Understanding Chemistry.* Holden-Day, Inc., San Francisco.
Taylor, J. C. (1976) Systematic features in the structural chemistry of the uranium halides, oxyhalides and related transition metal and lanthanide halides. *Coord. Chem. Rev.*, **20**, 197–273.

Chapter 2

SPECTROSCOPY

The purpose of this chapter is to outline the spectroscopic techniques which are most commonly applied to complexes of the f-block elements. These include electronic absorption, luminescence, NMR and EPR spectroscopies. Some applications of lanthanide luminescence and of magnetic resonance properties of lanthanide complexes are also described.

2.1 ELECTRONIC ABSORPTION SPECTROSCOPY

Three types of electronic transition can occur for lanthanide and actinide systems. These are: f→f transitions, nf→(n+1)d transitions, and ligand→metal f charge transfer transitions. This discussion will be limited to lanthanide complexes as these are much more straightforward than actinide systems, where increased spin-orbit coupling and larger crystal field effects result in more complex spectra.

The f–f transitions for lanthanides and actinides are, like d–d transitions of transition metal ions, electric-dipole forbidden by the Laporte selection rule. Interaction with the ligand field or with vibrational states mixes in electronic states with different parity and so f→f transitions become possible. However because of the small radial extent of the f-orbitals, these interactions are weak and the intensity of f→f transitions is therefore much lower than that of d→d transitions (a typical ε for a f→f transition is 5 $M^{-1}cm^{-1}$), and the absorptions are very much sharper. The colours of the Ln^{3+} ions are summarized in Table 2.1, and the electronic absorption spectrum of aqueous $PrCl_3$, along with the appropriate transitions, is shown in Figure 2.1.

The electronic spectrum of aqueous Ce^{3+} shows no absorptions in the visible region of the spectrum: the $^2F_{5/2}\rightarrow^2F_{7/2}$ transition occurs at low energy in the infrared region of the spectrum, where it is masked by vibrational transitions, and there is an intense 4f→5d transition in the ultraviolet.

It has been emphasized that there is very little interaction between f electrons and the crystal field, particularly for lanthanide complexes, but electronic spectra do show evidence of some interaction with the crystal field in the form of a shifting to lower frequencies of the absorptions of complexes compared with those of the free ions. This has been explained in terms of a nephelauxetic effect resulting from a small degree ($\leq$ 2.5% for lanthanides) of metal-ligand covalent bonding. Another crystal field effect is manifested in the form of hypersensitive transitions. The intensities of these transitions are found to be extremely sensitive to the ligand environment, varying by up to three orders of magnitude, depending on the nature of the ligands.

TABLE 2.1 ELECTRON CONFIGURATIONS, GROUND STATE TERM SYMBOLS AND COLOURS OF LN^{3+} IONS.

Ln	4f configuration	ground state term symbol	Colour of Ln^{3+}
La	$4f^0$	$^1S_0$	colourless
Ce^{3+}	$4f^1$	$^2F_{5/2}$	colourless
Pr^{3+}	$4f^2$	$^3H_4$	green
Nd^{3+}	$4f^3$	$^4I_{9/2}$	lilac
Pm^{3+}	$4f^4$	$^5I_4$	pink
Sm^{3+}	$4f^5$	$^6H_{5/2}$	pale yellow
Eu^{3+}	$4f^6$	$^7F_0$	colourless
Gd^{3+}	$4f^7$	$^8S_{7/2}$	colourless
Tb^{3+}	$4f^8$	$^7F_6$	very pale pink
Dy^{3+}	$4f^9$	$^6H_{15/2}$	pale yellow
Ho^{3+}	$4f^{10}$	$^5I_8$	yellow
Er^{3+}	$4f^{11}$	$^4I_{15/2}$	pink
Tm^{3+}	$4f^{12}$	$^3H_6$	pale green
Yb^{3+}	$4f^{13}$	$^2F_{7/2}$	colourless
Lu^{3+}	$4f^{14}$	$^1S_0$	colourless

Not all f→f transitions are hypersensitive: of the lanthanides Nd^{3+} and Er^{3+} show the greatest effects and Ce^{3+}, Gd^{3+} and Yb^{3+} show no hypersensitive transitions at all. The actinides have much larger crystal field effects than the lanthanides but Am^{3+} is the only actinide for which hypersensitive transitions have been observed. There is no conclusive explanation of these hypersensitive transitions, but the most dramatic effects are seen for low symmetry complexes and those with polarizable ligands. Most hypersensitive transitions have $\Delta J = \pm 2$. Henrie, Fellows and Choppin (1976), and Bridgeman and Gerloch (1997) have given accounts of the intensities of f→f transitions. Because of the dramatic intensity changes of hypersensitive transitions, they can be used to monitor complex formation in solution.

f→d transitions are Laporte allowed and therefore have much higher intensity than f→f transitions. They are also much broader, having a typical half-width of 1000 cm^{-1}. Ligand-to-metal charge transfer transitions are also Laporte allowed and therefore have high intensity. They are usually broader than f→d transitions; for easily reduced Ln^{3+} (Eu and Yb) the charge transfer transitions are at lower energy than the f→d transitions, and for easily oxidized ligands they may tail into the visible region of the spectrum, giving rise to much more intensely coloured complexes. For example the tris(silylamides) of Eu and Yb (Chapter 3) are quite intense orange and yellow respectively, whereas other $[Ln\{N(SiMe_3)_2\}_3]$ are very pale in colour like their parent Ln^{3+} ions.

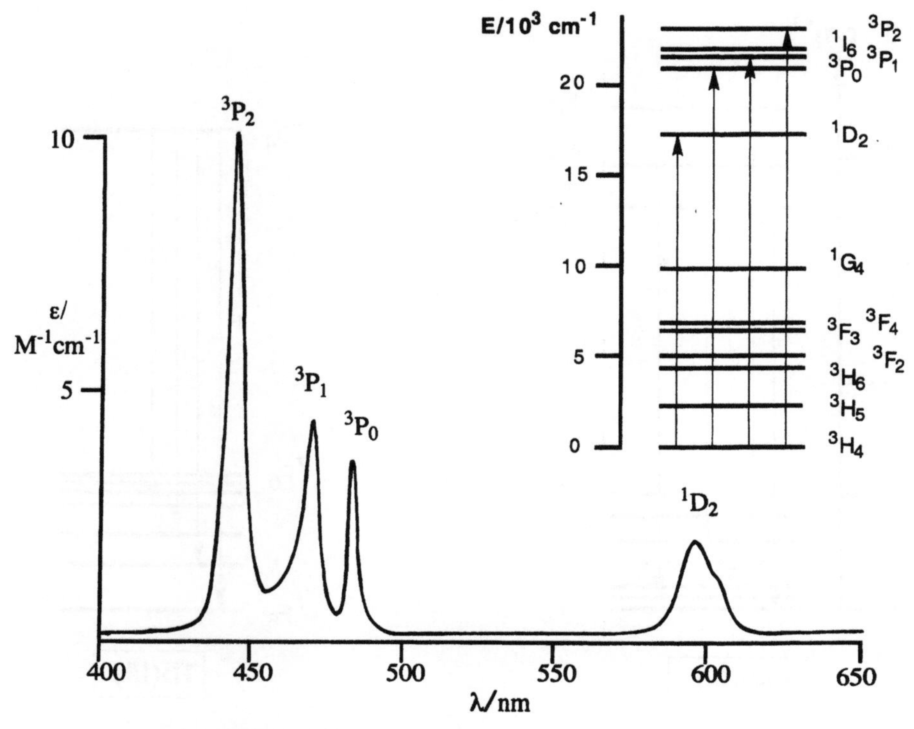

FIGURE 2.1 ELECTRONIC ABSORPTION SPECTRUM OF AQUEOUS $PrCl_3$.

- The interaction of f-orbitals with the crystal field is very small, especially for Ln complexes.
- f→f absorptions are sharp and have low extinction coefficients.
- Most f→f transitions are relatively insensitive to the nature of the ligands and the symmetry of the crystal field.
- Some 'hypersensitive' transitions (usually with ΔJ±2) are very sensitive to the symmetry of the crystal field.

2.2 LANTHANIDE LUMINESCENCE

The term 'luminescence' is used to describe a whole range of phenomena which involve decay from an electronically excited state by emission of a photon. Fluorescence and phosphorescence are the emission of photons after a sample has been excited by electromagnetic radiation; the distinction between the two is that fluorescence is a spin-allowed process taking 10^{-6} to 10^{-12} s whereas phosphorescence involves a change in spin multiplicity and is a slower process, taking from 10^{-6} s to as much as several seconds. The term 'luminescence' will be used here to describe both fluorescence and phosphorescence.

Most lanthanide ions luminesce in the solid state, and unlike luminescence from organic molecules, lanthanide emissions are sharp lines. This property has been used in lasers (*e.g.* the neodymium YAG laser) and in lanthanide phosphors (*e.g.* the red

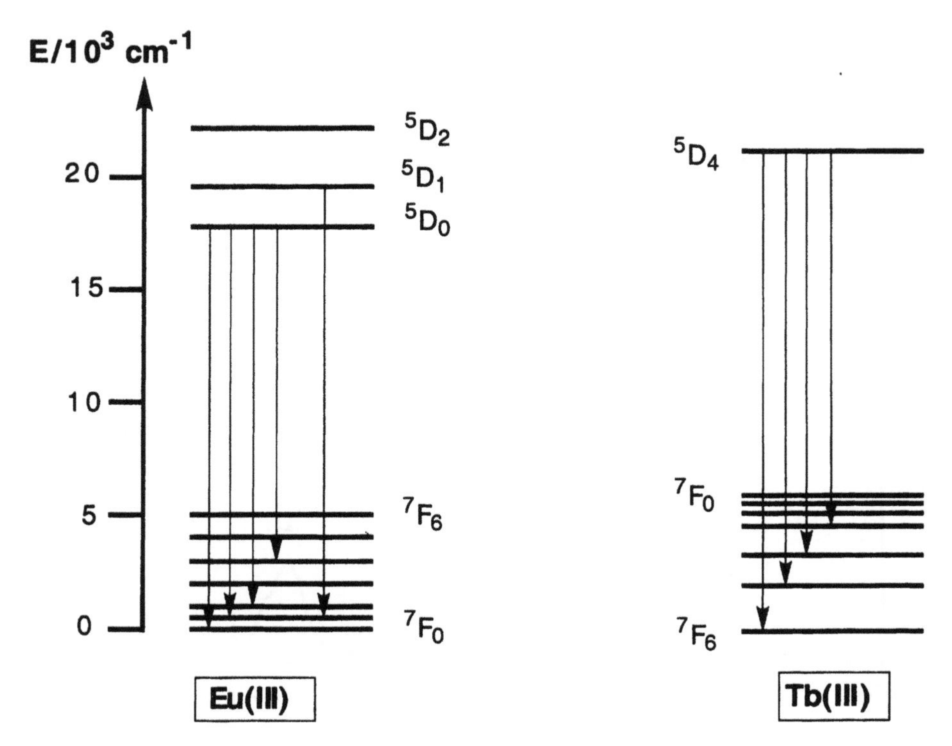

FIGURE 2.2 ENERGY LEVEL DIAGRAMS FOR EU^{3+} AND TB^{3+} LUMINESCENCE.

colour in TV's is provided by luminescence from Eu^{3+} doped into yttrium oxysulfide). Eu(III) and Tb(III) display the strongest luminescence (red for Eu(III) and green for Tb(III)), and this discussion will be limited to these two ions. The use of Eu(III) and Tb(III) as luminescent probes has been reviewed by Bünzli (1989) and by Horrocks and Albin (1984). Sabatini and Guardigli (1993) have reviewed the applications of luminescent lanthanide complexes as photochemical supramolecular devices.

The appropriate energy level diagrams for Eu^{3+} and Tb^{3+} luminescence are shown in Figure 2.2. f–f transitions are Laporte forbidden and so excitation of Ln^{3+} to an emissive state by this route is not an efficient process. However, once the ion has been excited to the emissive state (the most important states are $^5D_0$ for Eu^{3+} and $^5D_4$ for Tb^{3+}), luminescence can occur providing other non-radiative processes do not take over. The rate of non-radiative de-excitation is strongly related to the size of the energy gap between lowest emissive state and the highest state of the ground manifold; in the lanthanide series this gap is largest for Eu^{3+} (12 150 cm^{-1}), Gd^{3+} (32 000 cm^{-1}) and Tb^{3+} (14 800 cm^{-1}). This energy gap can be bridged by weak vibronic coupling of the excited state with high frequency oscillators such as an overtone of the ν(O-H) vibration of coordinated H_2O molecules. For Eu^{3+} the energy gap is bridged by three quanta of O-H vibrational energy as shown in Figure 2.3; four quanta are required to bridge the corresponding gap for Tb^{3+}. As a result of this facile de-excitation pathway, luminescence is not observed for aqueous solutions of Eu^{3+} or Tb^{3+}. Quenching of the excited states in D_2O solution is less

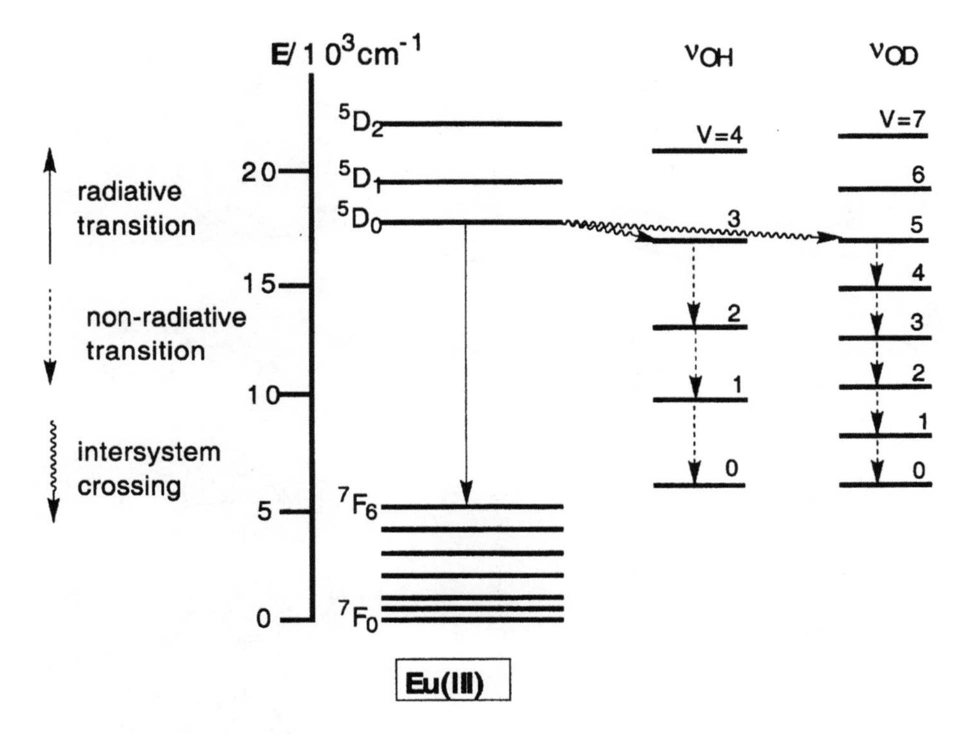

FIGURE 2.3 QUENCHING OF Eu^{3+} EMISSION BY ν(O-H).

effective than in H_2O: five quanta of the lower energy ν(O-D) are needed to bridge the energy gap for Eu^{3+} and six for Tb^{3+}. Efficient luminescence requires that the lanthanide ion is well separated from any high frequency oscillators.

The cryptand ligand 2.2.2 encapsulates lanthanide ions, reducing the number of coordinated H_2O molecules in aqueous solution to approximately two, and luminescence in aqueous solution is displayed by complexes of both Eu^{3+} and Tb^{3+}. The structure of $[Tb(2.2.2)(NO_3)]^{2+}$ is shown in Figure 2.4. Sabbatini, Dellonte and Blasse (1986) have investigated the photophysical properties of $[Ln \subset 2.2.2]^{3+}$ and found a quantum yield of 30% for emission from the Tb^{3+} complex. This is an order of magnitude greater than that from the Eu complex, and has been partly explained by the larger energy gap between the emissive state and the highest state of the ground manifold for Tb^{3+} compared with Eu^{3+}.

Sensitized luminescence

The efficiency of lanthanide luminescence can also be enhanced by increasing the efficiency of excitation. Because f–f transitions are Laporte forbidden, direct excitation to the emissive level cannot be an efficient process. An alternative method of excitation is *via* an organic ligand, usually a conjugated system, which has an excited triplet state higher in energy than the Ln^{3+} emissive state. On irradiation, the ligand molecule is excited into a singlet state, intersystem crossing then gives the ligand excited triplet state, and if the energy of this state is appropriate, fast intramolecular energy transfer to the lanthanide emissive state can occur as shown in Figure 2.5.

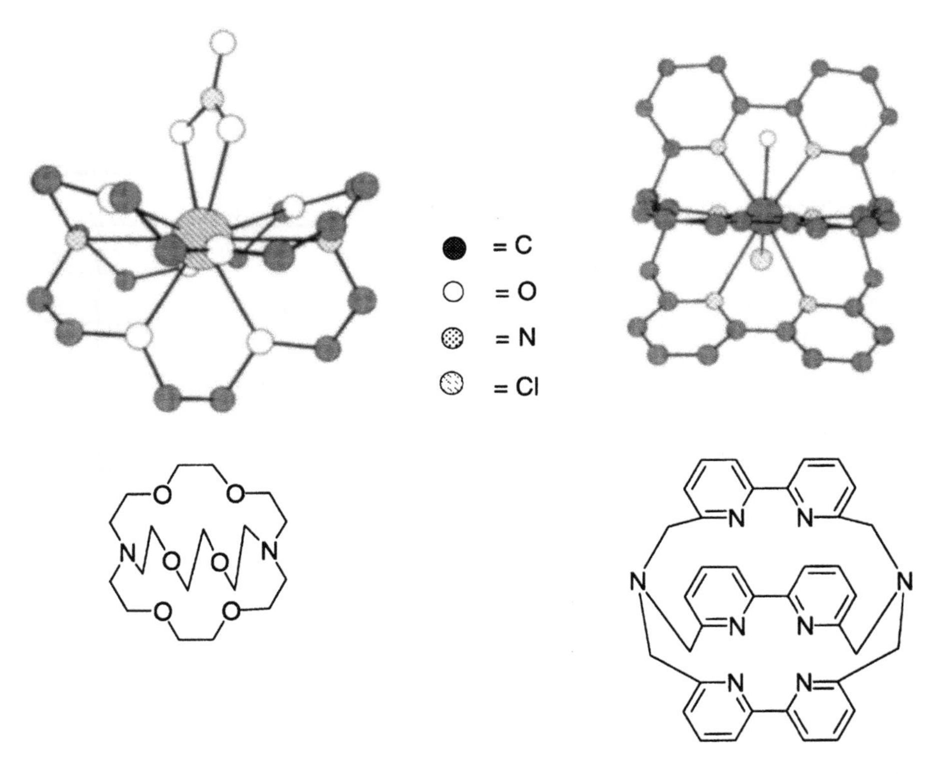

FIGURE 2.4 STRUCTURES OF $[Eu(2.2.2)(NO_3)]^{2+}$ AND $[Tb(bipy2.2.2)(H_2O)Cl]^{2+}$.

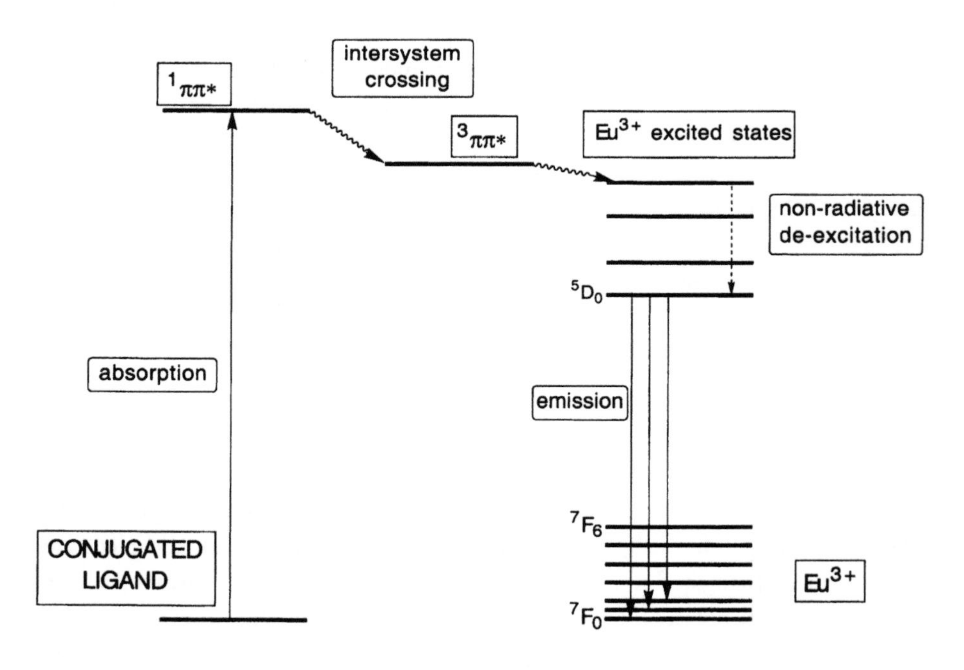

FIGURE 2.5 SENSITIZED LANTHANIDE LUMINESCENCE.

This phenomenon was first observed for lanthanide tris(β-diketonates), but much more spectacular results were reported by Alpha, Lehn and Mathis (1987) using macrobicyclic polypyridine ligands such as bipy2.2.2 shown in Figure 2.4. The excitation spectra of these polypyridine complexes are very similar to the absorption spectra of the free ligands, and the emission spectra are characteristic of Ln^{3+} luminescence.

- Lanthanide luminescence has technological applications *e.g.* Eu^{3+} in TV screens and Nd^{3+} in lasers.
- Eu^{3+} and Tb^{3+} show strong luminescence in the visible region which can sometimes be observed in solution.
- Luminescence is quenched in the presence of high-frequency oscillators such as ν(O-H) of coordinated H_2O.
- Sensitized Ln emission can occur *via* energy transfer from organic ligands with excited triplet states of appropriate energy.

2.2.1 Luminescence Spectroscopy

Luminescence spectroscopy includes excitation spectroscopy, where the intensity of a single emission is monitored while scanning excitation frequencies, and emission spectroscopy where the sample is irradiated with a fixed excitation frequency while the intensity and frequency of emitted photons is scanned. In this section we will consider emission spectroscopy, which is an extremely sensitive technique. Figure 2.2 shows the appropriate energy level diagrams for Eu^{3+} and Tb^{3+} luminescence, along with the most important transitions. For Eu^{3+} the most important emissive state is $^5D_0$ and for Tb^{3+} it is $^5D_4$; the properties of the most important transitions are summarized in Tables 2.2 and 2.3.

TABLE 2.2 PRINCIPAL TRANSITIONS IN THE EMISSION SPECTRUM OF Eu^{3+}.

Transition	Range	Intensity	Comments
$^5D_0 \rightarrow ^7F_0$	577–581	very weak	Forbidden Non-degenerate: if more than one transition appears then more than one Eu containing species is present
$^5D_0 \rightarrow ^7F_1$	585–600	strong	Allowed Intensity independent of environment Strong optical activity
$^5D_0 \rightarrow ^7F_2$	610–625	strong to very strong	Hypersensitive (ΔJ = 2) Absent if ion lies on inversion centre
$^5D_0 \rightarrow ^7F_3$	640–655	very weak	Forbidden Always very weak J mixing adds an allowed magnetic dipole character
$^5D_0 \rightarrow ^7F_4$	680–710	medium to strong	Sensitive to Eu environment (ΔJ = 4)
$^5D_1 \rightarrow ^7F_1$	530–540	very weak	Sensitive to Eu environment

TABLE 2.3 PRINCIPLE TRANSITIONS IN THE EMISSION SPECTRUM OF Tb^{3+}.

Transition	Range	Intensity	Comments
$^5D_4 \rightarrow {}^7F_6$	480–505	medium to strong	Sensitive to Tb environment
$^5D_4 \rightarrow {}^7F_5$	535–555	strong to very strong	Best transition for use as probe
$^5D_4 \rightarrow {}^7F_4$	580–600	medium to strong	Sensitive to Tb environment
$^5D_4 \rightarrow {}^7F_3$	615–625	medium	Displays strong optical activity
$^5D_4 \rightarrow {}^7F_2$	640–655	weak	Sensitive to Tb environment

The Eu^{3+} $^5D_0 \rightarrow {}^7F_0$ transition is a transition between two non-degenerate states which cannot be split however large the crystal field, and so can only give rise to a single line in the emission spectrum. If more than one line appears it indicates the presence of more than one Eu containing species. The $^5D_0 \rightarrow {}^7F_2$ transition is hypersensitive and is absent if the ion lies on an inversion centre. Because of the different quenching effects of D_2O and H_2O, measurements of luminescent lifetimes in D_2O and H_2O can allow estimation of the number of coordinated H_2O molecules in a complex. The luminescence spectrum of a macrocycle complex of Eu^{3+} is shown in Figure 2.6.

- Luminescence spectroscopy is a sensitive technique for studying complexes of Eu^{3+} and Tb^{3+}.
- Information about symmetry can be obtained from relative intensities of transitions.
- The number of coordinated H_2O ligands can be determined from measurement of luminescent lifetimes.

2.2.2 Applications of lanthanide luminescence in lighting and displays

An essential property of a good fluorescent lamp is good colour rendering. This is a measure of the ability to display an irradiated object in a natural way, and is quantified using the colour rendering index (CRI) which can take values from 0–100. A black-body radiator such as the sun or an incandescent lamp has a CRI of 100. It was demonstrated in the 1970's that good colour rendering can be achieved with just three narrow-band emitters in the blue, green and red regions of the spectrum. Lanthanide emitters are available for all of these frequencies: Eu^{2+} (450 nm), Tb^{3+} (540 nm) and Eu^{3+} (610 nm), and these are used in the so-called 'colour 80' lamps which have a CRI of 80–85 as well as a very high quantum efficiency ($\geq 90\%$). Although 'colour 80' lamps give excellent colour rendering for most purposes, they are often not so good for viewing lanthanide complexes, whose sharp absorption lines will not necessarily correspond to the sharp emissions of the lamp. Jüstel, Nikol and Ronda (1998) have reviewed the use of lanthanides in lighting and displays.

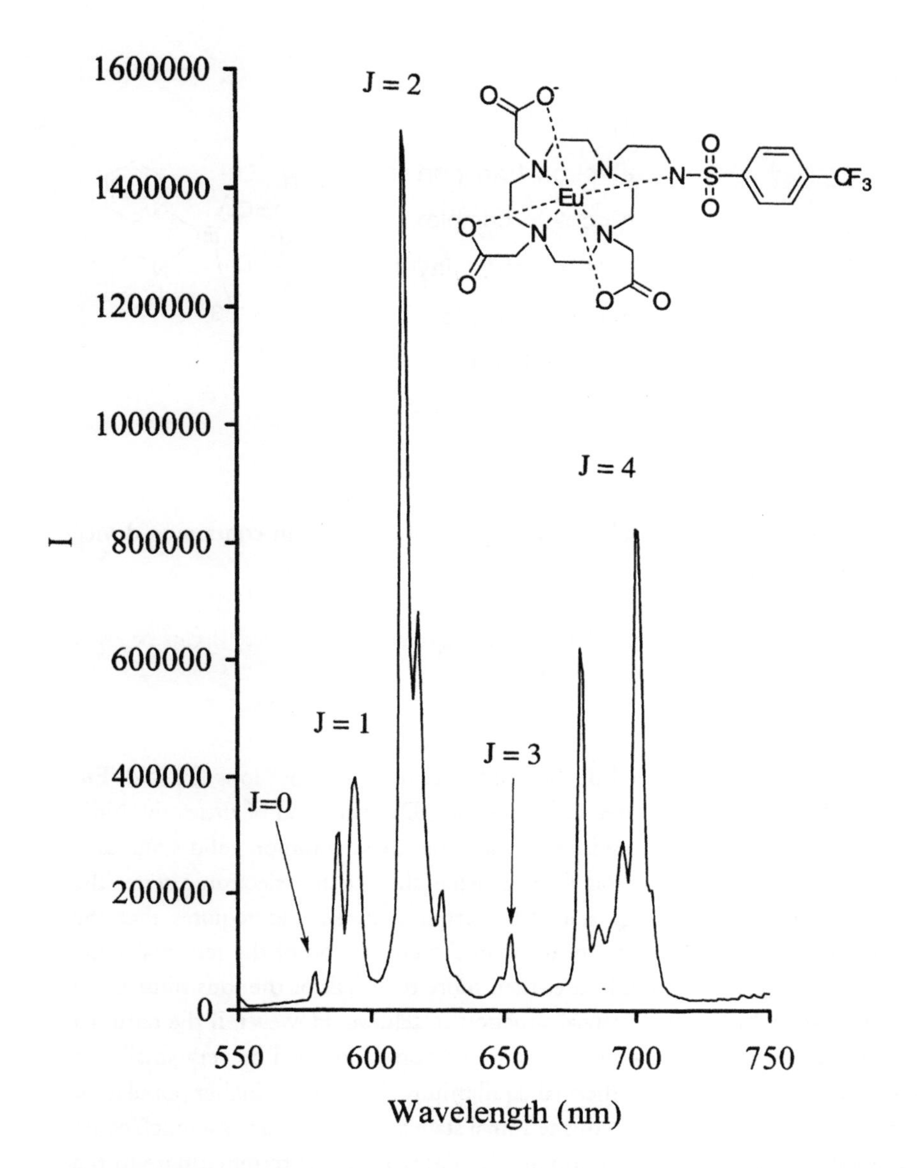

FIGURE 2.6 LUMINESCENCE SPECTRUM OF A COMPLEX OF Eu^{3+} WITH A SUBSTITUTED DOTA LIGAND (D. PARKER AND M.P. LOWE, UNIVERSITY OF DURHAM).

Electroluminescence occurs when charge carriers combine and give out light. Inorganic electroluminescence has been studied for many years, but the first electroluminescent device based on low molecular weight organic materials was reported in 1987. The sharp line emissions of Eu^{3+} and Tb^{3+} make them attractive components for the red and green emitters in a three-colour display. Europium tris(β-diketonates) are relatively stable complexes, soluble in organic solvents and are known to display luminescence in the solid state. A green-emitting electroluminescent device incorporating a terbium β-diketonate complex was first reported by Kido *et*

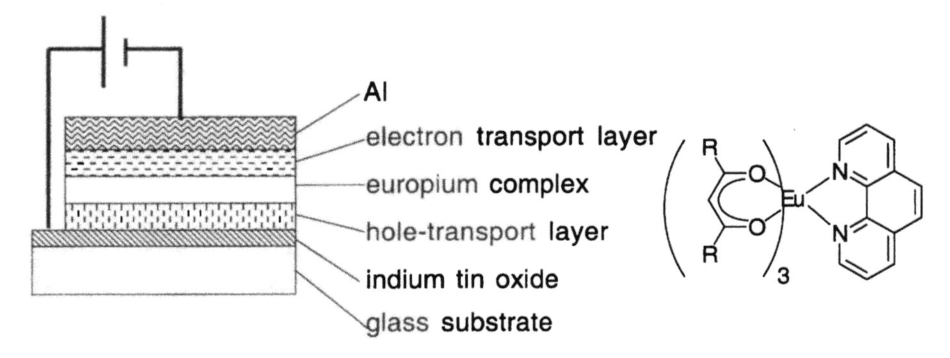

FIGURE 2.7 A LANTHANIDE ELECTROLUMINESCENT DEVICE.

al. (1990). Figure 2.7 shows a schematic diagram of a europium-containing device reported by Liu *et al.* (1997).

- Sharp line emission from Eu^{2+} (blue), Tb^{3+} (green) and Eu^{3+} (red) can be used in 3-colour lighting and displays

2.2.3 Neodymium lasers

Laser emission has been observed in the solid state for nine Ln^{3+} ions (Pr, Nd, Eu, Tb, Dy, Ho, Er, Tm, Yb) and three Ln^{2+} ions (Sm, Dy, Tm), and of these, the Nd^{3+} laser is by far the most widely used, in fact it is the most common solid state laser. Laser emission involves a transition from a metastable excited electronic state (the emissive state) to a lower energy state (the terminal state), and requires that the population of the emissive state is greater than the population of the terminal state. If the terminal state is the ground state, then more than half of the ions must be in the emissive state, which is not a trivial situation to achieve. However, if the terminal state has higher energy than the ground state, its population will be very small and determined just by the Boltzman thermal equilibrium. Achieving a higher population of the emissive state with respect to this terminal state therefore becomes a much easier task, especially if there is efficient non-radiative decay from the terminal state to the ground state. This is the principle of the four-level laser, and the Nd^{3+} ion has all the required properties. The wavelength of emission from the Nd^{3+} laser is 1.06 μm, in the infra red; a schematic energy level diagram is given in Figure 2.8.

Optical pumping excites the Nd^{3+} ions into the ^4F manifold. There is then fast non-radiative decay to the metastable $^4F_{3/2}$ state, which is the emisssive state, giving a population inversion between $^4F_{3/2}$ and $^4I_{11/2}$ (the terminal state). The $^4F_{3/2}\rightarrow{}^4I_{11/2}$ transition is the laser emission. The $^4I_{11/2}$ terminal level is approximately 2000 cm^{-1} above the $^4I_{9/2}$ ground state and at room temperature its fractional population is approximately 10^{-4}.

The most common Nd^{3+} laser is the Nd:YAG laser in which Nd^{3+} ions are doped at a concentration of approximately 1% into a host of yttrium aluminium garnet (YAG, $Y_3Al_5O_{12}$), giving a concentration of approximately 10^{20} Nd^{3+} atoms/cm^3.

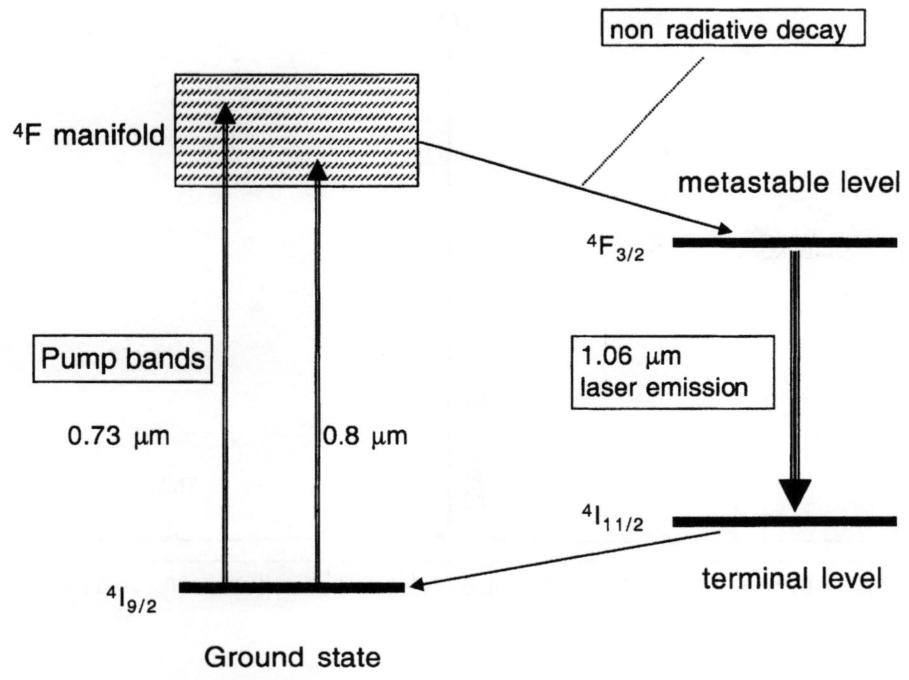

FIGURE 2.8 SCHEMATIC ENERGY LEVEL DIAGRAM FOR A Nd^{3+} LASER.

The Nd^{3+} ions absorb most strongly in the range 0.7 to 0.8 µm, and a variety of broadband lamps can be used to optically pump Nd lasers; these include tungsten, mercury, xenon and krypton. The area of rare earth lasers has been reviewed by Weber (1979); a practical account of Nd lasers has been given by Hecht (1986).

- The Nd-YAG laser is one of the most common solid state lasers.
- Its emission is in the infrared.
- It is an example of a 4-level laser.

2.3 NUCLEAR MAGNETIC RESONANCE

NMR spectroscopy of paramagnetic complexes, the applications of lanthanide shift reagents, NMR spectroscopy of lanthanide nuclei and the use of Gd^{3+} complexes as relaxation agents for NMR imaging will be considered in this Section.

2.3.1 NMR spectroscopy of paramagnetic complexes

Almost all complexes of lanthanides and actinides are paramagnetic due to the presence of unpaired f-electrons (although Eu^{3+} ($4f^6$) has a diamagnetic $^7F_0$ ground state it has paramagnetic excited states of sufficiently low energy to be significantly occupied at room temperature). When an NMR spectrum is recorded for a paramagnetic complex, the nucleus under investigation experiences the local magnetic field due to the paramagnetic metal ion in addition to the external magnetic field due to the NMR spectrometer. As a result the NMR spectra of paramagnetic

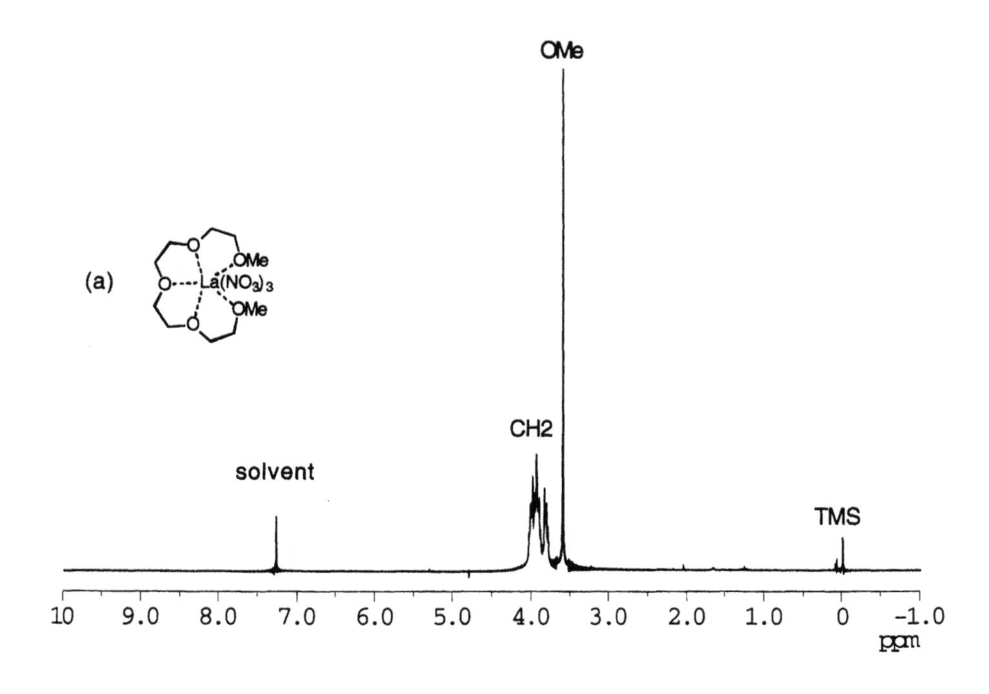

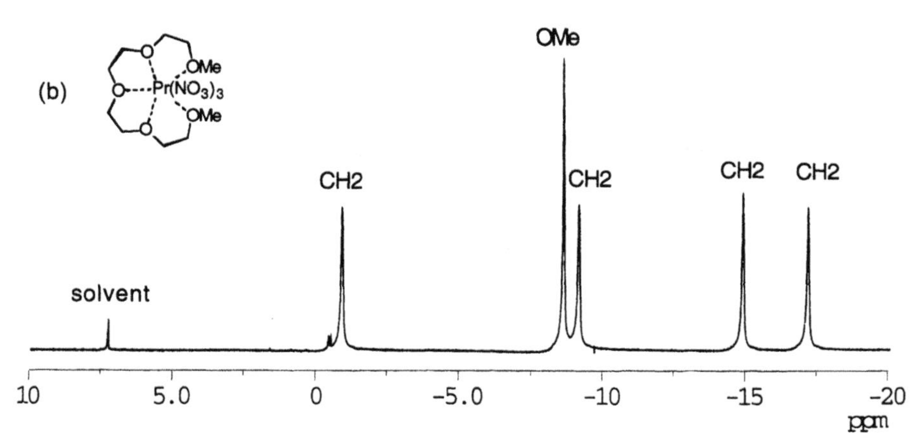

FIGURE 2.9 (A) ^{1}H NMR SPECTRUM OF DIAMAGNETIC [$La(NO_3)_3$(TETRAGLYME)] (B) ^{1}H NMR SPECTRUM OF PARAMAGNETIC [$Pr(NO_3)_3$(TETRAGLYME)].

complexes are significantly different from those of diamagnetic analogues, and much larger chemical shifts are observed. For most paramagnetic lanthanide and actinide complexes, strong spin-orbit coupling results in fast electron spin-lattice relaxation and so their NMR spectra show reasonably sharp lines. Figure 2.9 shows ^{1}H NMR spectra of diamagnetic [$La(NO_3)_3$(tetraglyme)] and paramagnetic [$Pr(NO_3)_3$ (tetraglyme)]. The structure of these complexes is described in Chapter 3.

For lanthanides and actinides there is very little covalent contribution to metal-to-ligand bonding and thus essentially no delocalization of unpaired f-electrons onto ligands. The mechanism of interaction between ligand nuclei and unpaired f-electrons is therefore almost entirely 'through-space' or dipolar in origin, and requires the metal ion to have an anisotropic distribution of f-electrons. It is often referred to as 'pseudocontact shifting'. If there is a small covalent contribution to bonding, some unpaired f-electron density will be delocalized onto ligand atoms and a 'contact' shift will result.

Pseudocontact shifting cannot occur if the complex has an isotropic magnetic susceptibility. Complexes of Gd^{3+} ($4f^7$), by virtue of their ^8S ground state, have isotropic magnetic susceptibility and so cannot give rise to dipolar (or pseudocontact) shifts, but can give rise to contact shifts. Because of slow electron spin relaxation in Gd^{3+}, NMR spectra of its complexes are usually extremely broad. Highly symmetrical complexes (T_d or O_h) of other ions also have isotropic magnetic susceptibility and so cannot give rise to dipolar shifts. A highly symmetrical paramagnetic complex whose NMR spectrum has provoked considerable debate is [Cp_4U]. This tetrahedral complex shows a large paramagnetic shift in its ^1H NMR spectrum: in C_6D_6 [Cp_4U] shows a singlet at 20.42 ppm compared with a chemical shift of 1.10 ppm for the diamagnetic [Cp_4Th] analogue. As a tetrahedral complex cannot give rise to dipolar shifting, this paramagnetic shift has been interpreted by von Ammon, Kanellakopoulos and Fischer (1970) as evidence for a significant covalent contribution to the bonding.

Bleaney (1972) has given a theoretical treatment of paramagnetic shifts in lanthanide complexes. For an axially symmetrical complex (one with an axis of order 3 or higher) the pseudocontact shift for a nucleus is given by:

$$\Delta v = \frac{D(3\cos^2\theta - 1)}{r^3}$$

where the constant D depends on $1/T^2$ and on the magnetic properties of the metal ion, and may be positive or negative depending on M. The distance r between the nucleus under observation and the Ln^{3+} ion, and the angle θ between the vector **r** and the principal symmetry axis of the complex, are defined in Figure 2.10, which also shows the variation of the term ($3\cos^2\theta - 1$) with θ.

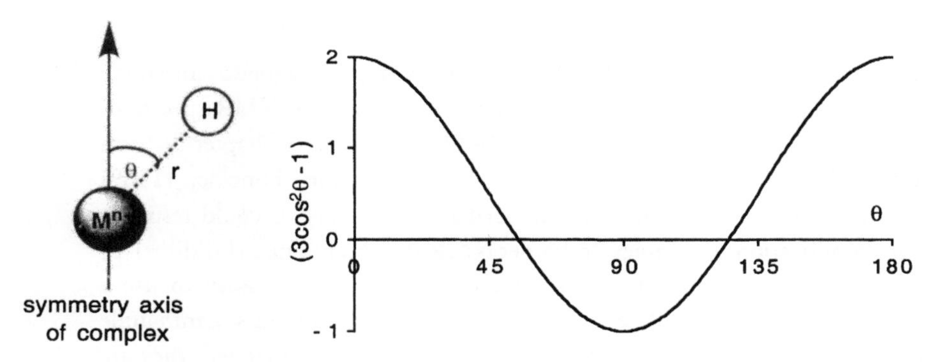

FIGURE 2.10 DEFINITIONS OF r AND θ TO DETERMINE THE PSEUDOCONTACT SHIFT FOR AN AXIALLY SYMMETRIC PARAMAGNETIC COMPLEX, AND THE VARIATION OF ($3\cos^2\theta - 1$) WITH θ.

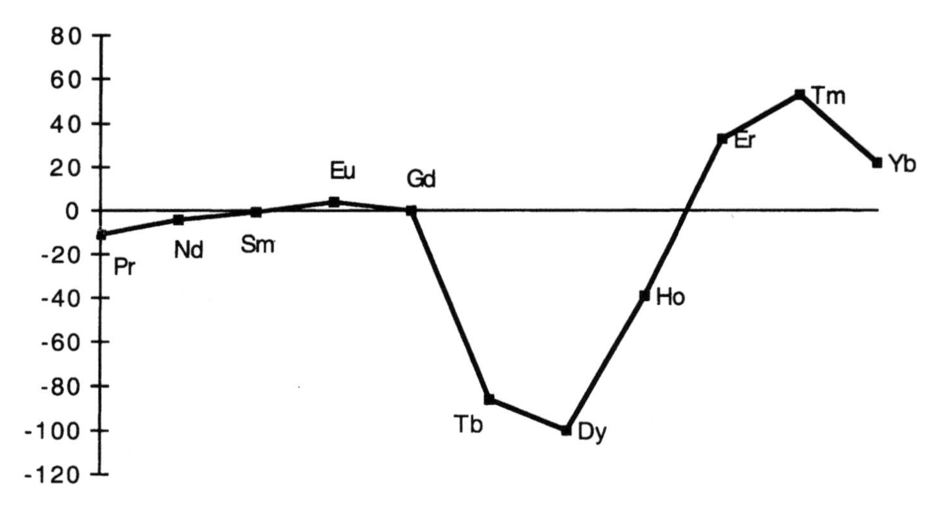

FIGURE 2.11 VARIATION OF RELATIVE DIPOLAR SHIFT WITH LN^{3+}. VALUES ARE SCALED TO −100 FOR DY^{3+}.

The magnitude and sign of $\Delta\nu$ are highly sensitive to θ as shown in Figure 2.10. When $\theta = 54.7°$ (half the tetrahedral angle) the term $(3\cos^2\theta - 1)$ is equal to zero and so there is no pseudocontact shift. The pseudocontact shift is clearly highly sensitive to geometry, depending on both r and θ, and under favourable circumstances can give a great deal of structural information. The other important factor in determining the magnitude and sign of $\Delta\nu$ is the magnetic properties of Ln^{3+}, and Figure 2.11 shows how the relative dipolar shift varies with Ln^{3+}.

- Most complexes of paramagnetic f-block ions display shifted NMR spectra.
- The shifting is almost entirely dipolar or 'pseudocontact' in origin.
- The sign and magnitude of $\Delta\nu$ depends on geometrical factors as well as on the magnetic properties of the metal ion.
- Paramagnetic S state ions (*e.g.* Gd^{3+}) do not give rise to dipolar shifting.

2.3.2 Lanthanide shift reagents

Historically much of the interest in NMR spectroscopy of paramagnetic lanthanide complexes has been due to their use as shift reagents to simplify NMR spectra of organic molecules. Anhydrous lanthanide tris(β-diketonates) (see Chapter 3) form labile adducts [Ln(diket)$_3$L] with Lewis bases such as alcohols. Hinckley (1969) recognized that the pseudocontact shift induced in coordinated L could result in significant simplification of complex 2nd order NMR spectra. Because [Ln(diket)$_3$L] are extremely labile with respect to exchange of L, it is not necessary to use a stoichiometric quantity of [Ln(diket)$_3$] in order to achieve good results. Lanthanide shift reagents have been referred to as 'the poor man's high-field magnet'; they are now much less used to simplify complex spectra because of the wide availability of high-field NMR spectrometers and 2-dimensional NMR spectroscopy.

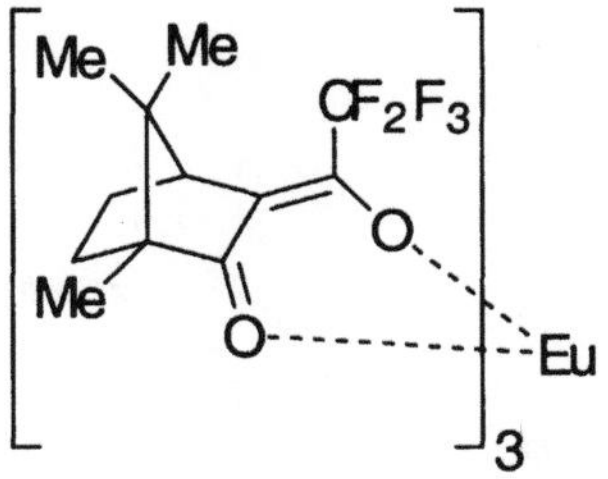

FIGURE 2.12 [EU(FACAM)$_3$].

Chiral lanthanide shift reagents

Determination of enantiomeric excesses is essential in asymmetric synthesis, and this is an area where lanthanide shift reagents are still valuable. The most common chiral LSR is [Eu(facam)$_3$] (Figure 2.12) where facam is a derivative of *d*-camphor.

This chiral complex is closely related to lanthanide tris(β-diketonates) and forms adducts with Lewis bases such as alcohols and amines. The *R* and *S* enantiomers in a racemic mixture of a chiral amine L cannot be distinguished by NMR spectroscopy. However in the presence of the chiral shift reagent [Eu(facam)$_3$] the adducts [Eu(facam)$_3$(*R*-L)] and [Eu(facam)$_3$(*S*-L)] will be formed. These adducts are diastereomers and so can be distinguished by NMR spectroscopy; they may also have different stability constants and be formed to different extents, resulting in different magnitudes of paramagnetic shift for *R*-L and *S*-L. The pseudocontact shift induced by the paramagnetic Eu^{3+} ion ensures that the chemical shifts are well separated and so enantiomeric excesses can be determined by integration of NMR resonances. Figure 2.13 shows ^{1}H NMR spectra of racemic α-methylbenzylamine in the absence and in the presence of the chiral shift reagent [Eu(facam)$_3$]. The resonances due to the amine are shifted downfield, away from the resonances due to the shift reagent which occur in the range –2.5 to +2.5 ppm.

Chiral lanthanide complexes have also been used in aqueous solution for analysis of water-soluble substrates such as amino acids. Kabuto and Sasaki (1984) used Eu^{3+} complexes of an EDTA analogue *R*-pdta and Hazama *et al.* (1996) used the polypyridine ligand *R*-tppn (Figure 2.14). [Eu(*R*-pdta)]$^-$ is expected to have a stability similar to that of [Eu(EDTA)]$^-$ for which $\log\beta = 17.32$, and substrate molecules bind in the sites normally occupied by H_2O. [Eu(*R*-tppn)]$^{3+}$ has a much lower stability in aqueous solution ($\log\beta = 5.6$ for the Pr^{3+} complex of an achiral analogue of tppn), however the complex is sufficiently stable to act as an effective chiral shift reagent.

Lanthanide shift reagents for use in biological systems

Lanthanide shift reagents have also been used to study the distribution of Na^+ and K^+ in biological systems. Both these ions are physiologically important, playing a crucial role in maintaining trans-membrane electrical potentials. ^{23}Na and ^{39}K are both 100% abundant quadrupolar nuclei, and are readily observable by NMR spectroscopy. However, the observed chemical shift is insensitive to the location of the M^+ ion (intacellular or extracellular) as it exists as aquo ions in both locations.

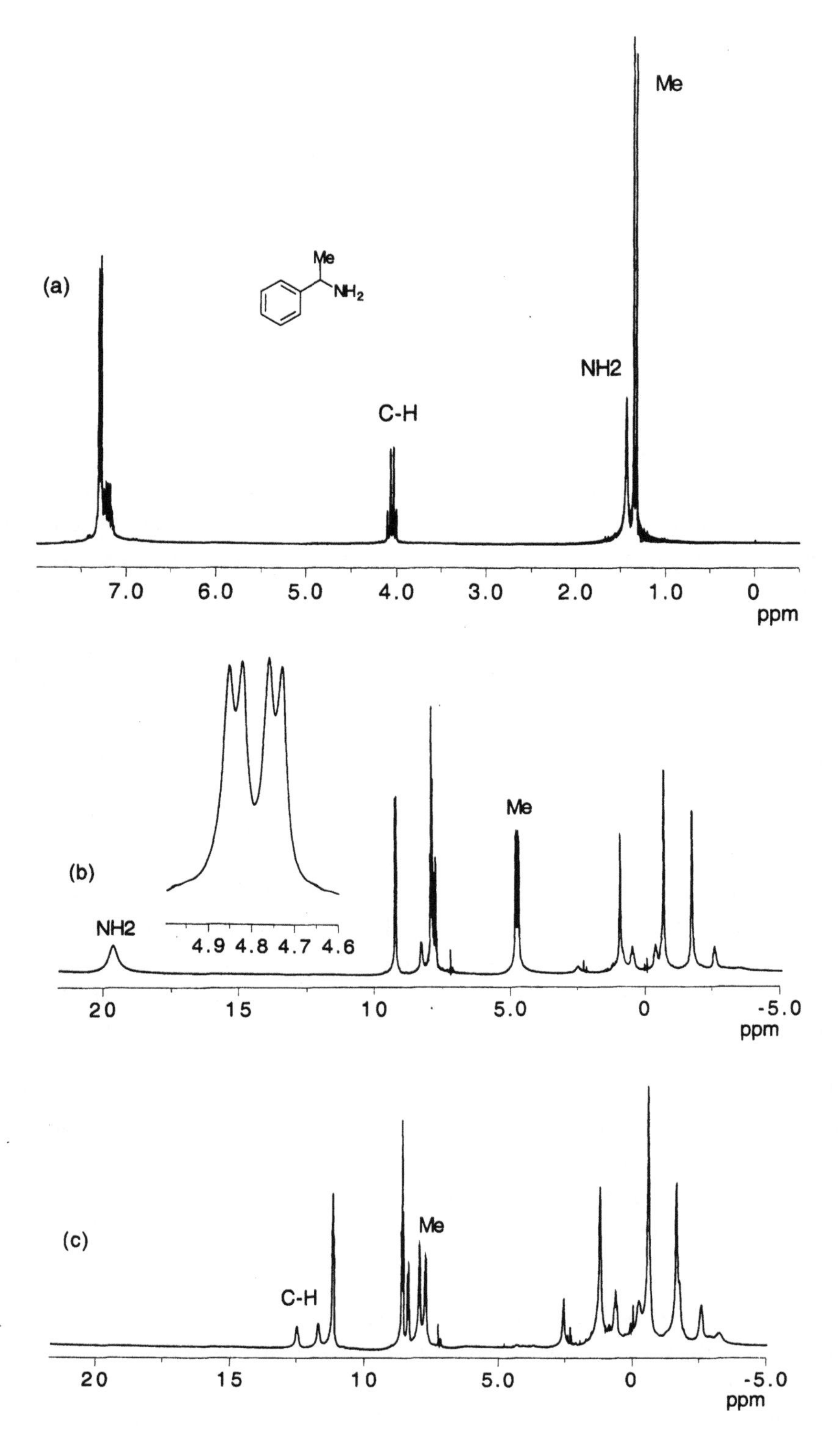

FIGURE 2.13 ^{1}H NMR SPECTRA SHOWING THE EFFECT OF A CHIRAL LANTHANIDE SHIFT REAGENT (A) RACEMIC α-METHYLBENZYLAMINE IN $CDCl_3$ (B) α-METHYLBENZYLAMINE (0.6M) + [EU(FACAM)$_3$] (0.16 EQ) (C) α-METHYLBENZYLAMINE (0.3M) + 0.33 EQUIV. [EU(FACAM)$_3$].

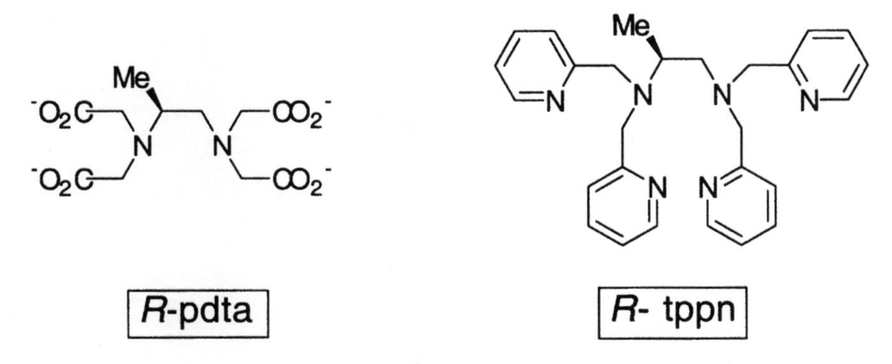

FIGURE 2.14 LIGANDS WHICH COORDINATE TO LN^{3+} FORMING CHIRAL SHIFT REAGENTS FOR USE IN AQUEOUS SOLUTION.

This problem has been overcome by the use of anionic lanthanide shift reagents which bind M^+ weakly, and which cannot enter the cell. These complexes remain in the extracellular space and shift only the extracellular M^+ ions. In this way NMR resonances due to inta- and extracellular M^+ can be separated and the relative concentrations of these species can be determined by integration. Lanthanide shift reagents for use in biological systems must be soluble and stable under physiological conditions (aqueous solution, pH 7) and they must be non-toxic. Examples of complexes which may be used as shift reagents are shown in Figure 2.15.

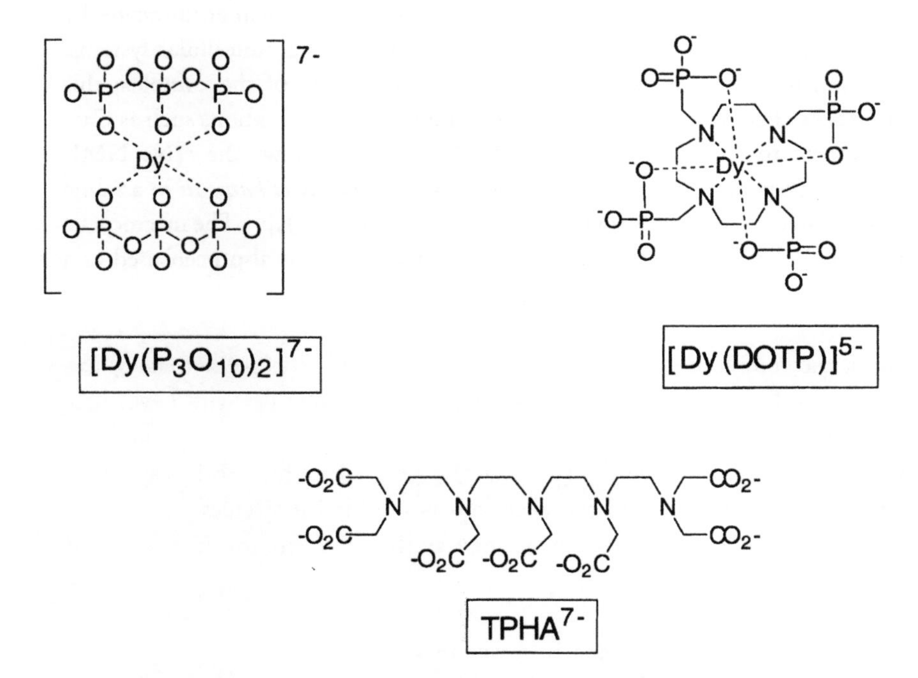

FIGURE 2.15 LANTHANIDE COMPLEXES USED AS NMR SHIFT REAGENTS IN BIOLOGICAL SYSTEMS.

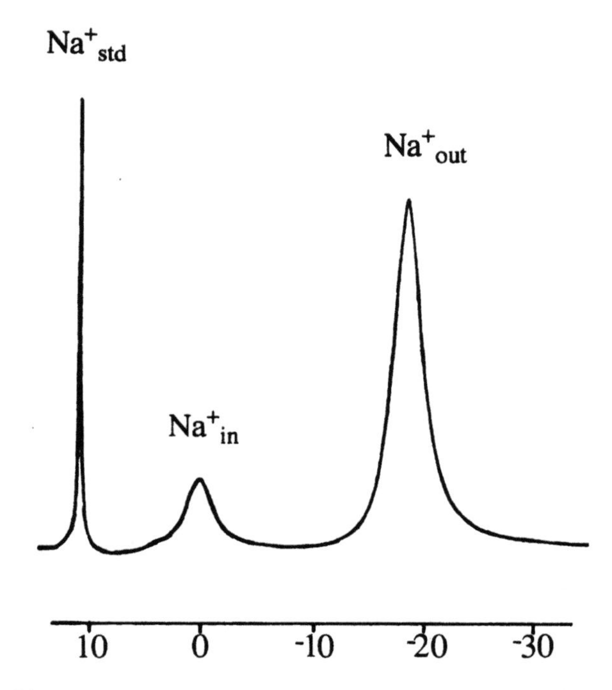

FIGURE 2.16 ^{23}NA NMR SPECTRUM OF THE SALT TOLERANT YEAST *DEBARYOMYCES HANSENII* IN A SALINE MEDIUM CONTAINING $Na_7[Dy(P_3O_{10})_2]$. (G. FROST, PHD THESIS, UNIVERSITY OF LIVERPOOL, 1991).

$[Dy(P_3O_{10})_2]^{7-}$ was used in the first ^{23}Na NMR study of human erythrocytes by Gupta and Gupta (1982) and has since been applied to several unicellular systems. M^+ ions bind reversibly to the negatively charged periphery of the complex; this reagent loses its effectiveness in the presence of multiply charged cations such as Ca^{2+} which compete for binding sites with M^+. Figure 2.16 shows the ^{23}Na NMR spectrum of a suspension of the salt tolerant yeast *Debaryomyces hansenii* in a saline growth medium containing the anionic shift reagent $[Dy(P_3O_{10})_2]^{7-}$. The macrocyclic ligand $DOTP^{8-}$ has high stability in aqueous solution and has also been used as a Na^+ shift reagent by Sherry *et al.* (1988).

- Lewis acidic complexes of Pr^{3+}, Eu^{3+} and Yb^{3+}, especially tris(β-diketonates) can be used as shift reagents to simplify NMR spectra of molecules with Lewis base properties.
- Chiral lanthanide complexes *e.g.* $[Eu(facam)_3]$ are used as chiral shift reagents to separate NMR signals due to the enantiomers of chiral molecules.
- Anionic lanthanide complexes can be used as shift reagents for *in vivo* NMR studies of alkali metals.

2.3.3 Gd^{3+} Complexes in Magnetic Resonance Imaging

Magnetic resonance imaging (usually referred to as MRI) is a technique of growing importance in diagnostic medicine. Edelman and Warach (1993) have given a concise review of the medical applications of MRI. The human body comprises at least 60% water, of which approximately 57% is intracellular and 43% is extracellular, and the magnetic resonance image is built up from NMR signals from H_2O molecules in the tissues. A two-dimensional image requires that the spatial location of each NMR signal is known; this is achieved by placing the patient in a magnetic field gradient so that H_2O molecules have a distinct resonant frequency depending upon their spatial location. MRI is particularly useful for soft-tissue imaging which cannot easily be achieved using other techniques. The intensity in a magnetic resonance image is dependent upon water proton relaxation times: protons with a short T_1 (longitudinal relaxation time) generally give rise to the highest signal intensities, which show up as bright areas in the image. As with other imaging techniques it is often necessary to enhance the contrast, for example in order to distinguish between healthy and diseased tissues. In MRI this is achieved by administering a paramagnetic contrast agent which localizes in a specific tissue and shortens T_1 for H_2O protons in that tissue.

Complexes of Gd^{3+} ($4f^7$) have a high magnetic moment, isotropic magnetic susceptibility, and slow electron spin relaxation rate, making them ideal candidates for use as MRI contrast agents. The dipolar interaction between the magnetic moment of Gd^{3+} and the proton spins of coordinated H_2O molecules will thus lead to reduction in T_1 for the H_2O, and an increase in image intensity.

The specific properties which must be addressed in the design of a contrast agent are: relaxivity, targeted distribution, solubility and stability, and lack of toxicity. Relaxivity R_i is a measure of the effectiveness of a relaxation agent. It is usually quoted in units of $mM^{-1}s^{-1}$ and is defined in the equation

$$(1/T_i)_{\text{observed}} = (1/T_i)_{\text{diamagnetic}} + R_i[\text{M}] \quad i = 1,2$$

where [M] is the concentration of the relaxation agent, $i = 1, 2$ refers to longitudinal (T_1) or transverse (T_2) relaxation times of H_2O. A plot of $(1/T_i)_{\text{observed}}$ against [M] gives a straight line with an intercept of $(1/T_i)_{\text{diamagnetic}}$ and a gradient of R_i. R_i is found to vary with frequency and temperature. The dipolar interaction has a $1/r^6$ dependence, falling off rapidly as the distance increases between the H_2O and the Gd^{3+} ion. Relaxation is therefore much more effective for inner-sphere (coordinated) than for outer-sphere H_2O molecules, and the ideal contrast agent will be able to accommodate several H_2O ligands which are in rapid exchange with bulk H_2O. Relaxivity is also found to be increased for complexes with long rotational correlation times τ_r so that a large molecule which tumbles slowly in solution will show shorter relaxation times than a small, freely tumbling spherical molecule. Table 2.4 gives relaxivities for Gd^{3+} species in aqueous solution.

Targeting of the contrast agent to specific tissues is not a trivial problem to solve, however the bio-distribution of a contrast agent is very dependent on its charge and structure, and a contrast agent usually shows enhanced activity in certain tissues even if it is not specifically located in those tissues. Charged complexes such as $[Gd(DTPA)]^{2-}$ are hydrophilic; they show minimal interaction with plasma proteins and cell

TABLE 2.4 RELAXIVITIES FOR Gd^{3+} COMPLEXES IN AQUEOUS SOLUTION. VALUES ALL DETERMINED AT 20 MHz.

Complex	no of H_2O ligands	$R_1/mM^{-1}s^{-1}$	T/°C	logK
Aquo ion[a]	9	9.1	35	
$[Gd(EDTA)]^-$ [a]	2–3	6.6	35	17.35
$[Gd(DTPA)]^{2-}$ [b]	1	4.3	25	22.46
[Gd(DTPA-BMA)][b]	1	4.39	25	16.85
$[Gd(DOTA)]^-$ [b]	1	4.8	40	25.30
[Gd(DO3A)][b]	1	4.8	40	21.0

[a] From Lauffer (1987)
[b] From Caravan *et al.* (1999)

membranes, and generally remain in the extracellular space. Tissues which contain the greatest fraction of extracellular space (*e.g.* tumours and abscesses) therefore show the greatest NMR signal enhancements, and $[Gd(DTPA)]^{2-}$ has been used with great success in the investigation of brain tumours. Figure 2.17 shows how administration of $[Gd(DTPA)]^{2-}$ contrast agent allows visualization of a lesion which is not visible in the absence of contrast agent or by X-ray mammography.

Contrast agents for *in vivo* use must of course be non-toxic. Gd^{3+} in the concentrations required for effective relaxation enhancement (*ca* 0.2 mmol per kg) is not well tolerated, and so contrast agents must be stable complexes to prevent release of Gd^{3+}, and must have adequate solubility under physiological conditions. Complexes of polyaminocarboxylates such as DTPA (See Chapter 3) were among the first to be used as they have high stability and are straightforward to prepare. However, charged species show increased osmotic effect compared with neutral molecules and the neutral Gd^{3+} complex with DTPA bis(amide) is better tolerated and is now used routinely under the name of Gadodiamide. LD_{50}(intravenous, mouse) is 14.8 mmol kg^{-1} for Gadodiamide compared with 5.6 mmol kg^{-1} for $Na_2[Gd(DTPA)]$. A selection of ligands used to complex Gd^{3+} in MRI contrast agents is shown in Figure 2.18.

For detailed articles on the use of Gd complexes as contrast agents for MRI see Lauffer (1987), Tweedle (1989), Aime *et al.* (1998) and Caravan *et al.* (1999).

- Contrast in MRI depends on relaxation times of protons in H_2O.
- Chelate complexes of Gd^{3+} ($4f^7$) are effective relaxation agents and are used as contrast agents in MRI.
- The most efficient relaxation is *via* an inner sphere mechanism *i.e.* the complex requires a binding site for H_2O.

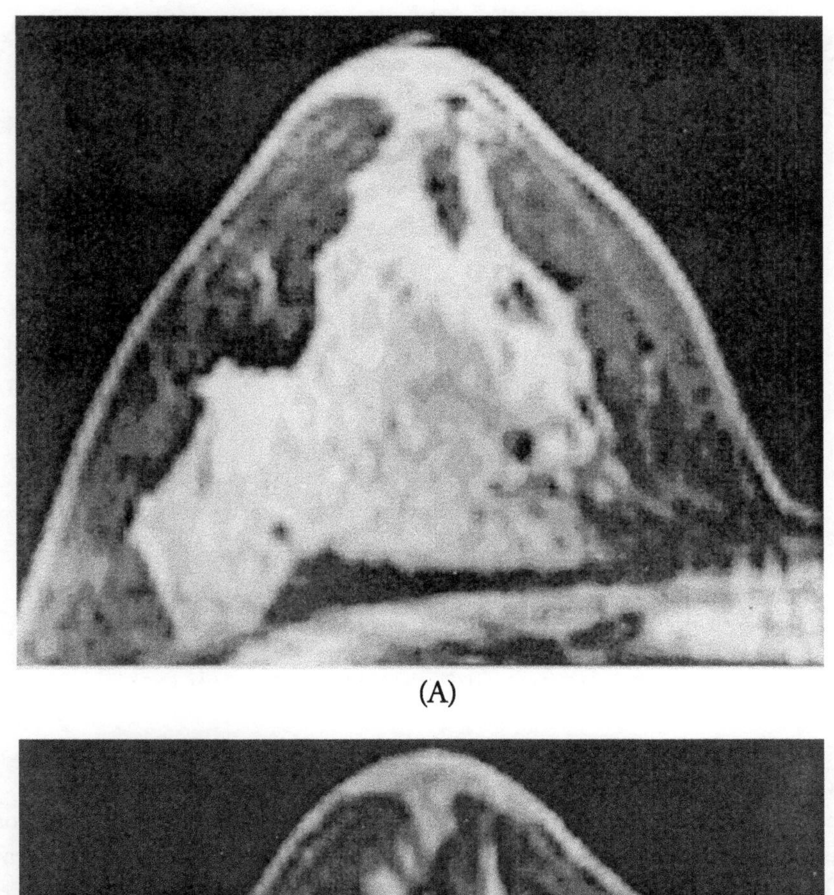

(A)

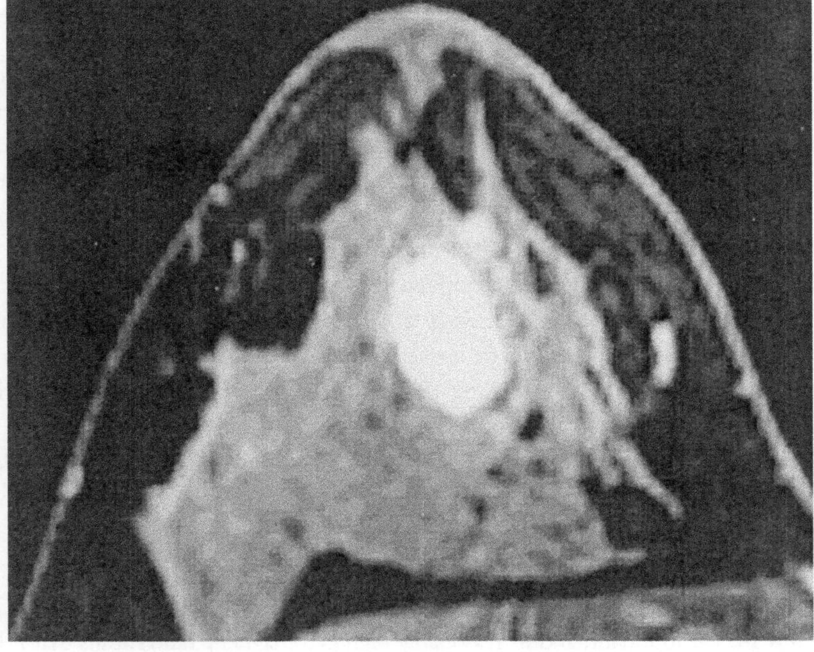

(B)

FIGURE 2.17 THE USE OF A MRI CONTRAST AGENT IN VISUALIZATION OF DISEASE. (A) MRI OF BREAST WITHOUT CONTRAST AGENT (B) MRI OF BREAST AFTER INJECTION OF 0.1 MMOL KG^{-1} $[Gd(DTPA)]^{2-}$, CLEARLY SHOWING FIBROADENOMA (BRIGHT AREA) WHICH IS NOT VISIBLE ON X-RAY MAMMOGRAPHY..(R. WEISSKOFF, EPIX MEDICAL, CAMBRIDGE MA.)

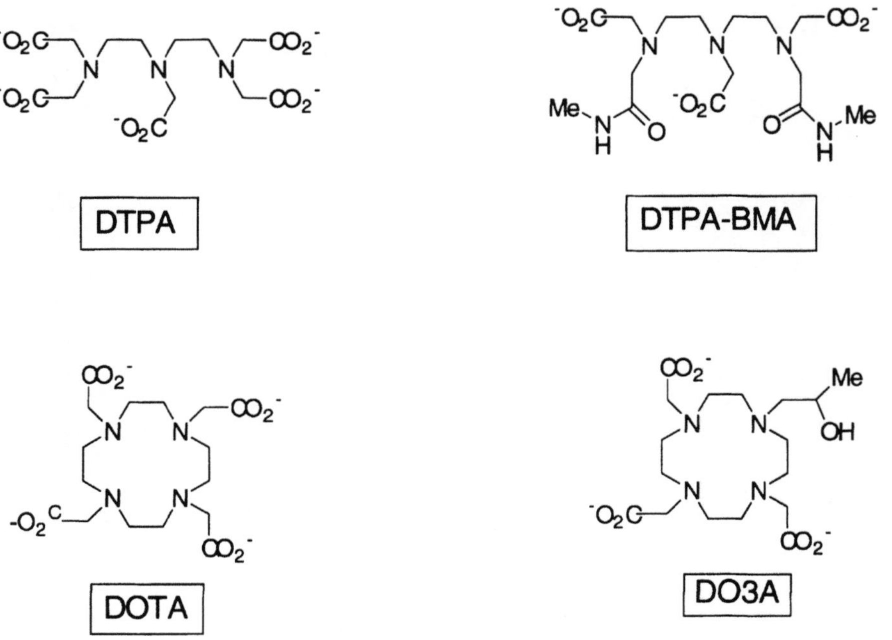

DTPA

DTPA-BMA

DOTA

DO3A

FIGURE 2.18 LIGANDS USED FOR Gd^{3+} IN MRI CONTRAST AGENTS.

2.3.4 NMR Spectroscopy of lanthanide nuclei

Direct NMR observation of metal nuclei can yield chemically useful information, but is only practical for diamagnetic complexes, and preferably those of elements with spin=1/2 nuclei, as quadrupolar nuclei often give unacceptably broad resonances. The only stable isotopes of rare earth or actinide elements with I=1/2 are ^{89}Y and ^{171}Yb, and together with the requirement for diamagnetic complexes this limits direct NMR observation of the metals to complexes of Y^{3+} or Yb^{2+}.

^{89}Y NMR spectroscopy was first applied to organometallic complexes by Evans *et al.* (1985). A series of complexes with η^5-C_5MeH_4 ligands was investigated and the ^{89}Y chemical shift was found to be very sensitive to the nature of the ligands as shown in Table 2.5. Coupling to H has been observed in yttrium hydride complexes, but longer range coupling (*e.g.* to the H atoms of the Me group in [$Cp'_2YMe(THF)$] is not observed. The disadvantages of ^{89}Y spectroscopy are the low observation frequency (9.8 MHz on a 200 MHz spectrometer), and the very long relaxation time, requiring a delay of 300s between 180° pulses. The long T_1 means that reasonable quality spectra can only be obtained in an acceptable time (*e.g.* overnight) on concentrated samples (*ca* 1 M).

^{171}Yb is 14.3% abundant, and its reasonably high gyromagnetic ratio means that it has a receptivity of 4.4 relative to ^{13}C. ^{171}Yb NMR was first applied to complexes of Yb(II) by Avent *et al.* (1989). A selection of ^{171}Yb chemical shifts (which are found to be very sensitive to temperature) is given in Table 2.6. Relaxation times for ^{171}Yb are much shorter (*ca* 1.3 s) than those for ^{89}Y, and ^{171}Yb spectra with acceptable signal to noise can be obtained in minutes rather than the many hours required for ^{89}Y NMR spectra.

TABLE 2.5 ^{89}Y NMR CHEMICAL SHIFTS OF ORGANOYTTRIUM COMPLEXES. (CP′ = C_5MeH_4) CHEMICAL SHIFTS ARE QUOTED IN PPM RELATIVE TO 3M AQUEOUS YCl_3.

Complex	^{89}Y chemical shift
$[Cp'_3Y(THF)]$	–371
$[Cp'_2YCl(THF)]$	–103
$[Cp'_2Y(\mu\text{-}H)(THF)]_2$	–92 (triplet, $^1J(Y\text{-}H)$ = 27 Hz)
$[Cp'_2Y(\mu\text{-}Me)]_2$	–15
$[Cp'_2YMe(THF)]$	+40

TABLE 2.6 ^{171}Yb CHEMICAL SHIFTS OF YB(II) COMPLEXES.

Complex	Temp/K	^{171}Yb chemical shift/ppm
$[(C_5Me_5)_2Yb(THF)_2]$	296	0
$[(C_5Me_5)_2Yb(OEt_2)]$	308	36
$[(C_5Me_5)_2Yb(py)_2]$	338	949
$[Yb\{N(SiMe_3)_2\}_2(OEt_2)_2]$	193	614
$[Yb\{N(SiMe_3)_2\}_2(Me_2PCH_2CH_2PMe_2)]$	193	1228

- ^{89}Y (100% abundant) and ^{171}Yb (14% abundant) are the only stable f-block isotopes with I=1/2.
- Direct observation is limited to diamagnetic complexes of Y^{3+} and Yb^{2+}.
- Relaxation times for ^{89}Y are long; those for ^{171}Yb are much shorter.

2.4 EPR SPECTROSCOPY

Most Ln and An ions have unpaired f electrons and so, in principle, EPR (electron paramagnetic resonance) spectra can be observed. In practice, because of large spin-orbit coupling, electronic relaxation is very rapid at room temperature for most of these ions and often it is only possible to observe sharp spectra at low temperatures (20K or less). Gd^{3+} ($4f^7$) is the exception: it has a ^8S ground state which, to a first approximation, has no orbital angular momentum and so cannot interact with a crystal field. This means that there is no facile mechanism for electron spin relaxation; at room temperature relaxation times for Gd^{3+} are of the order of 10^{-3} s and sharp EPR spectra can be observed.

In the free Gd^{3+} ion the M_J levels of the $^8S_{7/2}$ term are all degenerate, however in the presence of a crystal field, this term is split into four doublets (known as Kramers' doublets) with M_J = ±1/2, ±3/2, ±5/2 and ±7/2. The interaction of lanthanide ions with crystal fields is very small (typically of the order of 1 cm^{-1}) and so the Kramers' doublets are close in energy. When a magnetic field is applied, the degeneracy of the Kramers' doublets is lifted to give eight M_J levels as shown

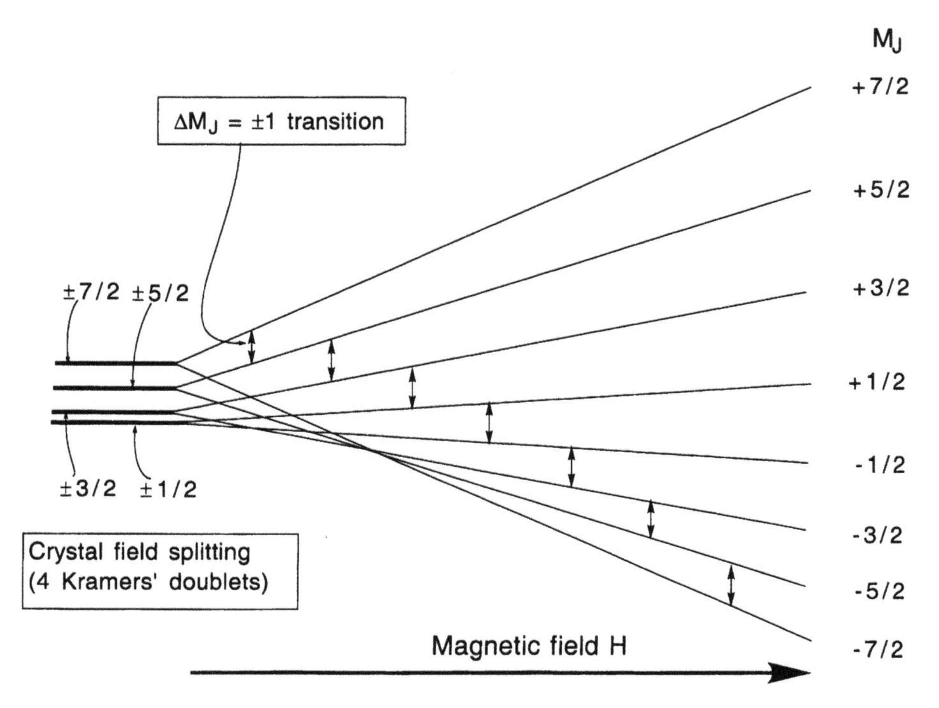

FIGURE 2.19 THE SPLITTING OF THE M_J LEVELS OF Gd^{3+} BY A CRYSTAL FIELD AND AN EXTERNAL MAGNETIC FIELD. THERE ARE SEVEN ALLOWED $\Delta M_J = \pm 1$ TRANSITIONS.

in Figure 2.19, where the magnitude of the splitting of each Kramers' doublet is proportional to the value of M_J. In EPR spectroscopy the sample is irradiated with a fixed frequency of microwave radiation, while increasing H (magnetic field strength). The allowed transitions in EPR spectroscopy have $\Delta M_J = \pm 1$. For Gd^{3+} there are seven allowed transitions, which are shown in Figure 2.19. It can be seen that the separation of the resonances in the EPR spectrum is dependent on the magnitude of the crystal field splitting: in the absence of a crystal field all the resonances would appear at the same value of H; when the crystal field is large, resonances will be seen at very low and very high values of H. The transition between $M_J = -1/2$ and $M_J = +1/2$ should always be observable and should have a *g*-value close to the free-electron value of 2.

Gd^{3+} is the f-block ion for which experimental EPR spectroscopy is the most straightforward. The energy of the microwave radiation used in EPR spectroscopy is of the same order of magnitude as crystal field splittings for Gd^{3+} complexes and so EPR spectroscopy is a very sensitive method for determining the details of crystal fields. In practice this requires the analysis of spectra of magnetically dilute single crystals in which a Gd^{3+} complex is doped at a level of $< 1\%$ into an isomorphous diamagnetic host (usually the analogous La or Y complex). Figure 2.20 shows the EPR spectrum of a single crystal of $[La(NO_3)_3(bipy)_2]$ doped with approximately 0.5% Gd^{3+}.

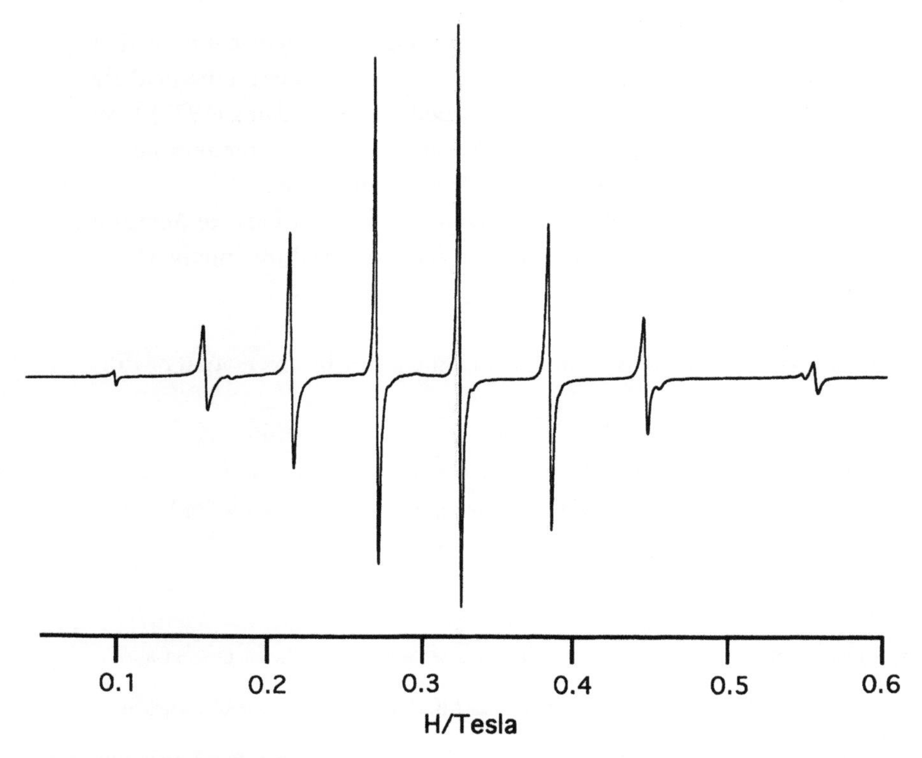

FIGURE 2.20 X-BAND EPR SPECTRUM OF A SINGLE CRYSTAL OF $[La(NO_3)_3(bipy)_2]$ DOPED WITH APPROXIMATELY 0.5% Gd^{3+}. THE CRYSTAL IS ORIENTED WITH CRYSTALLOGRAPHIC b AXIS PARALLEL TO THE MAGNETIC FIELD.

For any but the highest symmetry complexes a large number of parameters is required to describe the crystal field, and in order to evaluate these parameters a large amount of experimental data is required; these are usually obtained by recording spectra with the crystal in many orientations in the magnetic field. Most EPR studies of Gd^{3+} have been carried out on solid samples, and in order to extract the maximum amount of information, single crystals have been used to obtain information about crystal field strength and symmetry. Single crystal EPR spectroscopy has been used by Katoch and Sales (1980) to show that the crystal field splitting in the C_3 symmetric molecule $[Gd\{N(SiMe_3)_2\}_3]$ (Chapter 3) is exceptionally large. A full analysis of the EPR spectrum of the low symmetry (C_1) complex $[Gd(bipy)_2(NO_3)_3]$ has been carried out by Aspinall and Sales (1989).

Powder and frozen solution samples are of much less use than for transition metal complexes as superhyperfine coupling to ligand nuclei is almost never observed. However EPR spectroscopy is a very sensitive technique and Gd^{3+} has been used as a probe for Ca^{2+} in biological systems. Ca^{2+} is spectroscopically silent: it has no uv/vis spectrum, and is not NMR or EPR active. Gd^{3+} has a similar ionic radius to Ca^{2+} and many Ca^{2+} activated enzymes can also be activated by Gd^{3+}. For example, Musci, Reed and Berliner (1986) have investigated the Ca^{2+} binding site of α-lactalbumin using this technique. Although detailed and quantitative information cannot be extracted from such studies, observation of changes in spectra (particularly linewidth) can be useful. Reuben (1971) and Geraldes and Williams (1977) have measured linewidths in EPR spectra of aqueous solutions Gd^{3+} in the presence of a variety of ligands in order to assess the usefulness of these data.

For a very detailed account of EPR spectroscopy of 4f and 5f ions see Abragam and Bleaney (1970). Newman and Urban (1975) have reviewed the interpretation of EPR spectra of *S*-state ions.

- Most paramagnetic f-block ions require low temperature for observation of EPR spectra.
- EPR spectra of Gd^{3+} ($4f^7$) can be observed at room temperature.
- Superhyperfine coupling to ligand nuclei is generally not observed.
- Information about crystal field magnitude and symmetry is available from single crystal EPR spectra.

REFERENCES

Abragam, A. and Bleaney, B. (1970) *Electron Paramagnetic Resonance of Transition Ions*. London: Oxford University Press.

Aime, S., Botta, M., Fasano, M. and Terreno, E. (1998) Lanthanide (III) chelates for NMR biomedical applications. *Chem. Soc. Rev.*, 27, 19–29.

Alpha, B., Lehn, J.-M. and Mathis, G. (1987) Energy transfer luminescence of europium(III) and terbium(III) cryptates of macrobicyclic polypyridine ligands. *Angew. Chem. Int. Ed. Engl.*, **26**, 266–267.

Aspinall, H.C. and Sales, K.D. A single crystal ESR study of $Gd(bipy)_2(NO_3)_3$ in $La(bipy)_2(NO_3)_3$. *Spectrochim. Acta*, **45A**, 739–745.

Avent, A.G., Edelman, M.A., Lappert, M.F. and Lawless, G.A. (1989) the first high-resolution direct NMR observation of an f-block element. *J. Am. Chem. Soc.*, **111**, 3423–3425.

Bleaney, B. (1972) Nuclear magnetic resonance shifts in solution due to lanthanide ions. *J. Mag. Res.*, **8**, 91–100.

Bridgeman, A.J. and Gerloch, M. (1997) Spectral intensities of transition metal complexes. *Coord. Chem. Rev.*, **165**, 315–446.

Bünzli, J.-C.G. (1989) Luminescent probes. In *Lanthanide Probes in Life, Chemical and Earth Sciences*, edited by J.-C.G. Bünzli and G.R. Choppin, pp 219–293. Amsterdam: Elsevier.

Caravan, P., Ellison, J.J., McMurry, T.J. and Lauffer, R.B. (1999) Gadolinium(III) chelates as MRI contrast agents: structure, dynamics, and applications. *Chem. Rev.*, **99**, 2293–2352.

Edelman, R.R. and Warach, S. (1993) Magnetic resonance imaging. *New England J. Med.*, **328**, 708–716.

Evans, W.J., Meadows, J.H., Kostka, A.G. and Closs, G.I. (1985) Y-89 NMR-spectra of organoyttrium complexes. *Organometallics*, 4, 324–326.

Geraldes, E.C.N.F. and Williams, R.J.P. (1977) An investigation of some potential uses of the gadolinium(III) ion as a structural probe. *J. Chem. Soc., Dalton Trans.*, 1721–1726.

Gupta, R.K. and Gupta, P. (1982) Direct observation of resolved resonances from intra-cellular and extracellular Na-23 ions in NMR-studies of intact-cells and tissues using dysprosium(III)tripolyphosphate as paramagnetic shift-reagent *J. Mag. Res.*, 47, 344–350.

Hart, F.A. and Laming, F.P. (1965) Lanthanide complexes-II. Complexes of 1:10-phenanthroliine with lanthanide acetates and nitrates. *J. Inorg. Nucl. Chem.*, 27, 1605–1609.

Hazama, R., Umakoshi, K., Kabuto, C., Kabuto, K. and Sasaki, Y. (1996) A europium(III)-N,N,N',N'-tetrakis(2-pyridylmethyl)-(*R*)-propylenediamine complex as a new chiral lanthanide NMR shift reagent for aqueous neutral solution. *J. Chem. Soc. Chem. Commun.*, 15–16.

Hecht, J. (1986) *The laser guidebook*, pp 284–302. New York: McGraw-Hill.

Henrie, D.E., Fellows, R.L. and Choppin, G.R. (1976) Hypersensitivity in the electronic transitions of lanthanide and actinide complexes. *Coord. Chem. Rev.*, **18**, 199–224.

Hinckley, C.C. (1969) Paramagnetic shifts in solutions of cholesterol and the dypyridine adduct of trisdipivalomethanatoeuropium(III). A shift reagent. *J. Am. Chem. Soc.*, **91**, 5160–5162.

Horrocks, W. deW. and Albin, M. (1984) Lanthanide ion luminescence in coordination chemistry and biochemistry. *Prog. in Inorg. Chem.*, **31**, 1–104.

Judd, B.R. (1962) Optical absorption intensities of rare earth ions. *Phys. Rev.*, **127**, 750–761.

Jüstel, T., Nikol, H. and Ronda, C. (1998) New developments in the field of luminescent materials for lighting and displays. *Angew. Chem. Int. Ed.*, **37**, 3084–3103.

Kabuto, K. and Sasaki, Y. (1984) The europium(III) (*R*)-propylenediaminetetra-acetate ion: a promising chiral shift reagent for ^{1}H NMR spectroscopy in aqueous solution. *J. Chem. Soc. Chem. Commun.*, 316–318.

Katoch, D.S. and Sales, K.D. (1980) Single-crystal electron spin resonance studies of gadolinium trisilylamide and sodium gadolinium dipicolinate (pyridine-2,6-dicarboxylate). *J. Chem. Soc., Dalton Trans.*, 2476–2479.

Kido, J., Nagai, K. and Ohashi, Y. (1990) Electroluminescence in a Terbium Complex. *Chemistry Letters*, 657–660.

Lauffer, R.B. (1987) Paramagnetic metal complexes as water proton relaxation agents for NMR imaging: theory and design. *Chem. Rev.*, **87**, 901–907.

Liu, L., Hong, Z., Peng, J., Liu, X., Liang, C., Liu, Z., Yu, J. and Zhao, D. (1997) Europium complexes as emitters in organic electroluminescent devices. *Synthetic Metals*, **91**, 267–269.

Musci, G., Reed, G.H. and Berliner, L.J. (1986) An electron-paramagnetic resonance study of bovine alpha-lactalbumin-metal ion complexes.*J. Inorg. Biochem.*, **26**, 229–236.

Newman, D.J. and Urban, W. (1975) Interpretation of *S*-state ion E. P. R. spectra. *Adv. Phys.*, **24**, 793–844.

Ofelt, G.S. (1962) Intensities of crystal spectra of rare earth ions. *J. Chem. Phys.*, **37**, 511–520.

Reuben, J. (1971) Electron spin relaxation in aqueous solutions of gadolinium(III). Aquo, cacodylate and bovine serum albumin measurements. *J. Phys. Chem.*, **75**, 3164–3167.

Sabbatini, N., Dellonte, S. and Blasse, G. (1986) The luminescence of the rare earth cryptates $[Tb \subset 2.2.1]^{3+}$ and $[Sm \subset 2.2.1]^{3+}$. *Chem. Phys. Lett.*, **129**, 541–545.

Sabbatini, N., and Guardigli, M. (1993) Luminescent lanthanide complexes as photochemical supramolecular devices. *Coord. Chem. Rev.*, **123**, 201–228.

Sherry, A.D., Malloy, C.R., Jeffrey, F.M.H., Cacheris, W.P. and Geraldes, C.F.G.C. (1988) $Dy(DOTP)^{5-}$: A new stable ^{23}Na NMR shift reagent. *J. Mag. Res.*, **76**, 528–533.

Stein, G. and Würzberg, E. (1975) Energy gap law in the solvent isotope effect on radiationless transitions of rare earth ions. *J. Chem. Phys.*, **62**, 208–213.

Tweedle, M.F. (1989) Relaxation agents in NMR imaging. In *Lanthanide Probes in Life, Chemical and Earth Sciences*, edited by J.-C.G. Bünzli and G.R. Choppin, pp. 127–179. Amsterdam: Elsevier.

Von Ammon, R., Kanellakopoulos, B. and Fischer, R.D. (1970) NMR evidence for electron spin delocalization in organometallic U(IV) compounds. *Chem. Phys. Lett.*, **4**, 553–557.

Weber, M.J. (1979) Rare earth lasers. In *Handbook on the Physics and Chemistry of Rare Earths*, edited by K.A. Gschneidner and L. Eyring, Volume 4, Chapter 35. Amsterdam: North-Holland Publishing Company.

Chapter 3

COORDINATION CHEMISTRY

Lanthanide and actinide ions are hard Lewis acids and their interaction with ligands is almost exclusively electrostatic in nature (although in some actinide complexes there is evidence of a small degree of covalency). The f-orbitals are not involved to a significant extent in metal-to-ligand bonding and so, in contrast to d-transition metal complexes, bonds to ligands are not strongly directional in nature. The coordination geometry of lanthanide and actinide complexes is dictated almost exclusively by steric considerations: the ligands pack around the metal ions in such a way as to minimise interligand repulsions, and the coordination number is determined by the steric bulk of the ligands. Because of the large sizes of lanthanide and actinide ions, coordination numbers are usually high (coordination numbers of 8 or 9 are very common, and several complexes are known with coordination numbers of 12), and coordination geometries are often irregular. Lower coordination numbers can be achieved with very bulky ligands such as hexamethyldisilylamide $^{-}N(SiMe_3)_2$, whereas the highest coordination numbers are usually achieved with chelating ligands which have small bite-angles such as NO_3^-. One of the best examples of 12-coordination is the $[Ln(NO_3)_6]^{3-}$ ion which is shown in Figure 3.1.

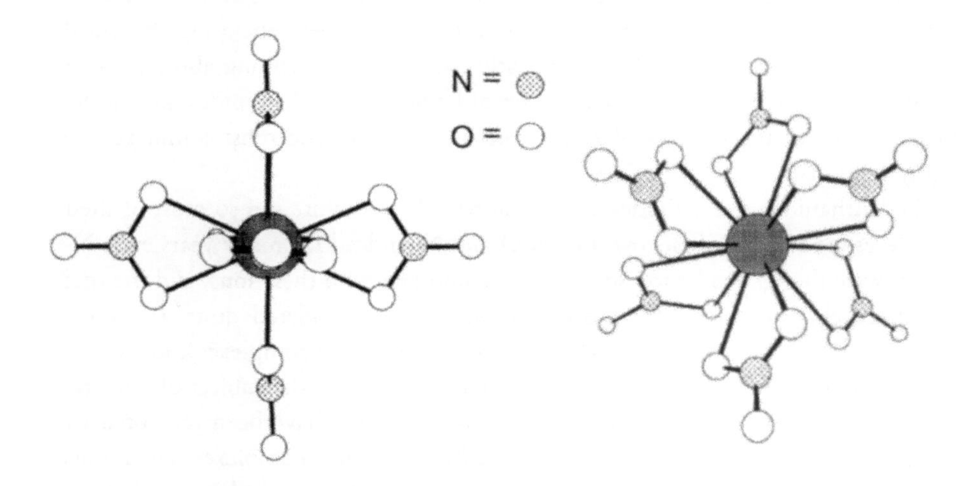

FIGURE 3.1 STRUCTURE OF $[La(NO_3)_6]^{3-}$.

TABLE 3.1 COORDINATION GEOMETRIES FOR LANTHANIDE AND ACTINIDE COMPLEXES.

Coordination number	Coordination geometry	Example for Ln	Example for An
3	Trigonal pyramidal	$[La\{N(SiMe_3)_2\}_3]$	$[U\{CH(SiMe_3)_2\}_3]$
4	Tetrahedral	$[LuBu^t_4]^-$	$[U(O\text{-}2,6\text{-}Bu^t_2C_6H_3)_4]$
5	Trigonal bipyramid		$[U(NEt_2)_4]_2$
5	Distorted trigonal bipyramid	$[Ce(O\text{-}2,6\text{-}Bu^t_2C_6H_3)_3(Bu^tNC)_2]$	
6	Octahedron	$[LnCl_6]^{3-}$	UF_6
6	Trigonal prism	$[Lu(thd)_3]$	
7	Pentagonal bipyramid		$[UO_2(NCS)_5]^{3-}$
7	Capped trigonal prism	$[Ho(thd)_3(H_2O)]$	
8	Square antiprism	$[Ce(acac)_4]$	$[Th(acac)_4]$
8	Cube	$[La(bipy\text{-}N\text{-}oxide)_4]^{3+}$	$[NEt_4]_4[U(NCS)_8]$
9	Tri-capped trigonal prism	$[Ln(H_2O)_9]^{3+}$	$[U(dipicolinate)_3]^{2-}$
10	Irregular	$[Ln(NO_3)_3(12\text{-}C\text{-}4)]$	$[Th(NO_3)_4(OPPh_3)_2]$
10	Bi-capped square antiprism		$[Th(dipicolinate)_2(H_2O)_4]$
11	Irregular	$[Ln(NO_3)_3(15\text{-}C\text{-}5)]$	$[Th(NO_3)_4(H_2O)_3]$
12	Icosahedron	$[Ln(NO_3)_6]^{3-}$	$[Th(NO_3)_6]^{2-}$
14	Bi-capped hexagonal antiprism		$[U(BH_4)_4]$

A selection of complexes with coordination numbers 3 to 12 is given in Table 3.1, along with the coordination geometries; many of these complexes will be encountered later in this Chapter. Bagnall and Xing-Fu (1982) have quantified the packing of ligands around lanthanide and actinide by using a cone-packing model. In this model the solid angles subtended by the ligands at the metal are summed, and this sum is found to be very close to $4\pi \times 0.8$ for complexes of both lanthanides and actinides, indicating that 80% of the surface of a sphere enclosing the metal ion is occupied by ligands. The electrostatic nature of the bonding also results in a high degree of lability in most complexes of lanthanides and actinides: kinetically stable complexes are only available with chelating ligands, and most complexes are highly fluxional.

The lanthanides and actinides are typical hard Lewis acids, and so most of their complexes are with the first-row donors O and N. Indeed for many years even N-donors were thought to be too soft to form complexes with these ions, and the first complexes characterized with N-donor ligands were considered quite a novelty. Although O and N donors predominate, there are now many complexes known with P and S donors, and an extensive organometallic chemistry (the subject of Chapter 4) has been developed. Sm-X bond disruption enthalpies have been measured by Nolan, Stern and Marks (1989) for a series of $[(C_5H_5)_2SmX]$ complexes. The results of these measurements are shown in Figure 3.2 from which it can be seen that the Sm-S bond is by no means weak.

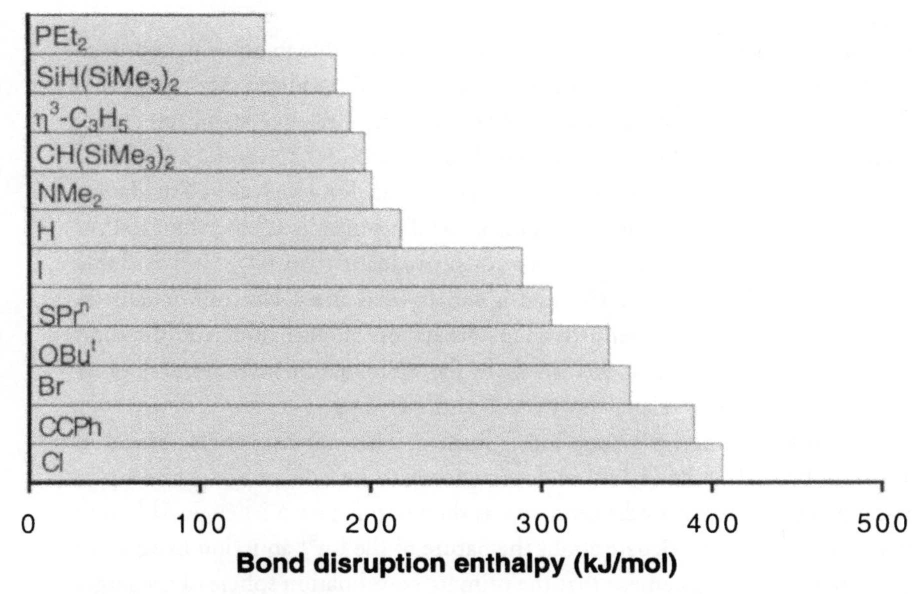

FIGURE 3.2 BOND DISRUPTION ENTHALPIES ($kJmol^{-1}$) FOR $[(C_5Me_5)_2Sm\text{-}X]$. SEE NOLAN, STERN AND MARKS (1989).

The aim of this Chapter is to give an account of the main features of lanthanide and actinide coordination chemistry, illustrated with a limited number of examples, many of which will be referred to in other Chapters. Comprehensive accounts of coordination chemistry are available elsewhere.

- Lanthanide and actinide ions are typical hard Lewis acids, forming most complexes with hard O or N donors.
- Large ionic radii lead to high coordination numbers.
- Bonding is ionic and non-directional, often leading to irregular coordination geometries.

3.1 COORDINATION CHEMISTRY OF LANTHANIDE AND ACTINIDE IONS IN AQUEOUS SOLUTION

It will be seen in this chapter that aqueous solution is not always the easiest medium in which to carry out lanthanide or actinide coordination chemistry. However, a knowledge of the behaviour of these ions in aqueous solution is essential to an understanding of their environmental chemistry (particularly important for the actinides), and there are a growing number of applications of lanthanides in physiology and medicine, all of which involve aqueous solutions around pH 7. These include contrast agents for magnetic resonance imaging, highly sensitive reagents for fluoroimmunoassay and radiolanthanide complexes for cancer therapy. Design of complexes which are stable under physiological conditions is an important and challenging goal.

3.1.1 Lanthanides

For all the lanthanides the +3 oxidation state is the most stable in aqueous solution. Eu^{2+} may be prepared by reduction with Zn, but it is slowly oxidized; Yb^{2+} and Sm^{2+} may also be prepared but are oxidized very rapidly. These relative stabilities reflect the E° values for the Ln^{3+}/Ln^{2+} couples which are –1.55, –0.35, and –1.05 V for Sm, Eu and Yb respectively. Cerium is the only lanthanide for which the +4 oxidation state is of importance in aqueous solution. In acidic solution E° for the Ce(IV)/Ce(III) couple is +1.72 V, and Ce^{4+} is a stronger oxidant than Cl_2. Ce^{4+} is stable in aqueous solution for several weeks, and is widely used as a 1-electron oxidant in both synthetic and analytical chemistry. This discussion of lanthanide coordination chemistry will be limited almost exclusively to the +3 oxidation state because of its overwhelming importance relative to the +2 and +4 states.

In aqueous solution the Ln^{3+} aquo ion is formed. This was first characterized in the solid state by Helmholz (1939) and was shown to be a 9-coordinate complex with tri-capped trigonal pyramidal geometry as shown in Figure 3.3 below. Although there has been considerable debate about the nature of the Ln^{3+} aquo ion in aqueous solution there is general agreement that the primary coordination sphere of the larger Ln^{3+} ions (La – Nd) contains between 9.0 and 9.3 H_2O molecules, and that of the smaller Ln^{3+} (Dy – Lu) contains between 7.5 and 8.0 H_2O molecules. A discontinuity in stability constant data which often occurs around Gd^{3+} has been ascribed to this change in hydration number around the middle of the lanthanide series.

Ln^{3+} ions undergo hydrolysis at pH values above about 6, and the species $Ln(OH)^{2+}$, $Ln_2(OH)_2{}^{4+}$, $Ln_3(OH)_5{}^{4+}$ have been reported. In neutral solution the concentrations of Ln^{3+} and $Ln(OH)^{2+}$ are approximately equal; in the absence of strongly coordinating ligands at pH ≥ 7, extensive hydrolysis occurs, and the hydroxy species are absorbed onto glass and suspended particles. At even higher pH's, hydroxy colloids are formed and eventually the highly insoluble (*ca* 2×10^{-4} g dm^{-3}) $Ln(OH)_3$ precipitate out of solution. Even before precipitation of hydroxide is actually observed, a significant loss of Ln^{3+} from solution may have occurred. Ce^{4+}, because of its higher charge and smaller ionic radius, undergoes hydrolysis at lower pH's.

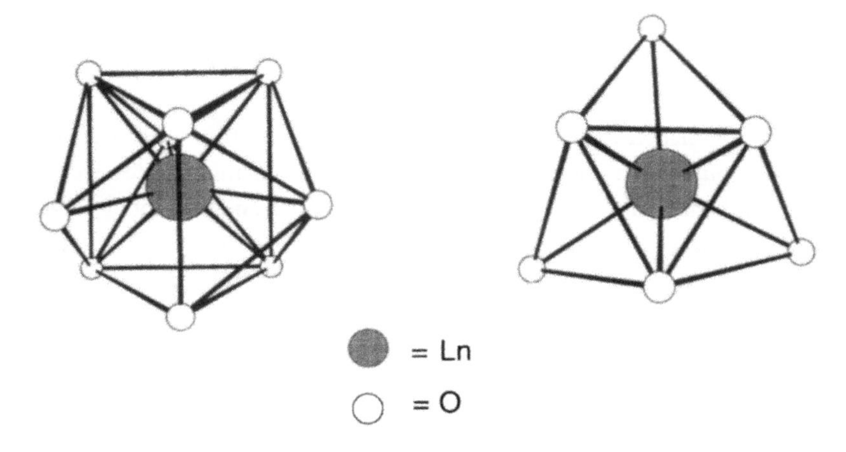

FIGURE 3.3 STRUCTURE OF $[Ln(H_2O)_9]^{3+}$.

- For all Ln the +3 oxidation state is the most stable in aqueous solution
- Eu^{2+} is oxidized slowly in aqueous solution; Sm^{2+} and Yb^{2+} are rapidly oxidized. Ce^{4+} is stable in aqueous solution for several weeks.
- Hydrolysis of Ln^{3+} occurs above pH6. Hydrolysis of Ce^{4+} occurs at lower pH.

3.1.2 Actinides

Unlike the lanthanides, the actinides can exist in a range of oxidation states in aqueous solution. These were summarized in Chapter 1.

Oxidation state +2

This is the most important oxidation state only for No, where the $5f^{14}$ configuration of No^{2+} leads to extra stability. The +2 oxidation state is not known for other An in aqueous solution.

Oxidation state +3

In aqueous solution this oxidation state is only of importance for Ac and the later actinides Am to Md, and Lr. The chemistry of these An^{3+} ions is very similar to that of the corresponding Ln^{3+} ions.

Oxidation state +4

In aqueous solution this is the only stable oxidation state for Th, and is one of two stable oxidation states for U. It is moderately stable for Pa and Np. Hydrolysis of the An^{4+} occurs between pH 2 and pH 3 and dinuclear species such as $[Th_2(OH)_2]^{6+}$ have been identified as well as polynuclear species such as $[Th_6(OH)_{14}]^{10+}$. At pH > 4 these highly charged ions are fully hydrolyzed and irreversible precipitation of AnO_2 occurs.

Oxidation state +5

This oxidation state is stable only for Pa and Np, which exist in aqueous solution as AnO_2^+. This linear cation is similar to the actinyl ion AnO_2^{2+} (see below) but is rather more flexible. Hydrolysis gives $AnO(OH)^{2+}$. The +5 oxidation state is not stable in aqueous solution for U, and UO_2^+ disproportionates to give U^{4+} and UO_2^{2+}. The An(V) oxidation state can be stabilized by addition of F^- which results in the formation of the $[AnF_6]^-$ ion.

Oxidation state +6

This oxidation state is known for U, Np, Pu and Am. In aqueous solution An(VI) exist as the actinyl ion AnO_2^{2+} whose structure is described below. The relative stability of the actinyl ions is U>Np>Pu>Am. The actinyl ion is a very persistent species and, particularly for UO_2^{2+} there is an extensive coordination chemistry.

The AnO_2^{2+} ions have short An-O distances (between 1.7 and 1.9 Å) and a linear O-An-O geometry, in contrast to the MoO_2^{2+} and WO_2^{2+} ions which have O-M-O angles of *ca* 110°, and the gas phase ThO_2 molecule which is also non-linear (O-Th-O = 122°). There has been much debate in the literature on the nature of bonding in AnO_2^{2+} and Denning (1992) has reviewed the area. It is generally accepted that there are significant covalent contributions to the bonding in these

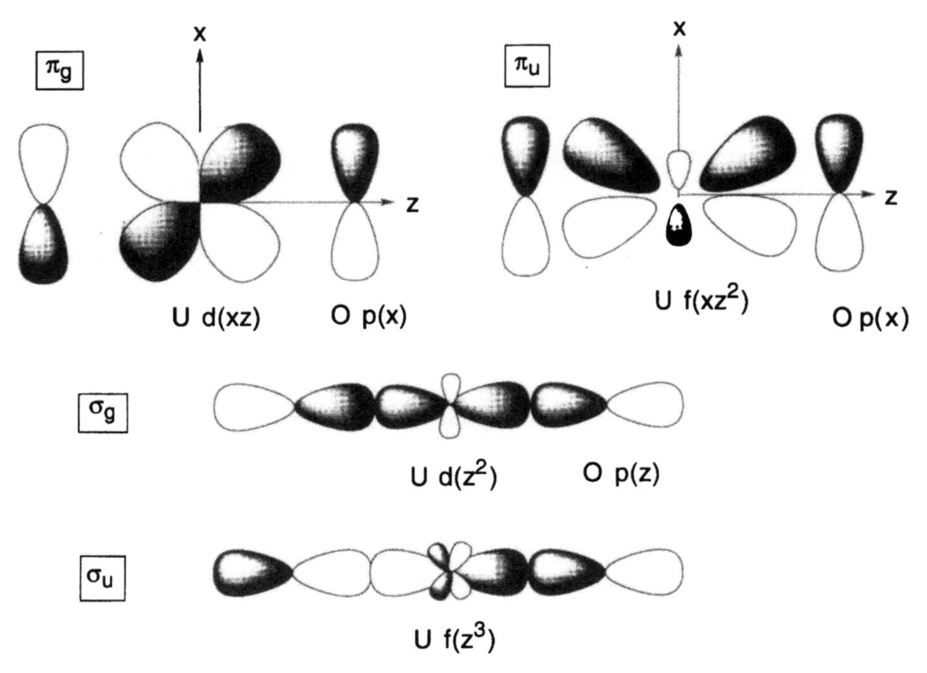

FIGURE 3.4 MOLECULAR ORBITALS FOR AnO_2^{2+}.

ions, accounting for their stability and the short An-O bond distances. A diagram showing O p-orbital interactions with d(xz), d(z^2), f(xz^2) and f(z^3) orbitals is shown in Figure 3.4.

Oxidation state +7

This oxidation state is known for Np, Pu and Am, but is rather unstable in solution for Pu and Am. In aqueous solution these ions exist as AnO_5^{3-} and are obtained by oxidation of An(VI) in alkaline solution by ozone, peroxides, XeF_2 or other strong oxidizing agents. Even Np(VII), the most stable, oxidizes both H_2O and Ce^{3+}.

- An ions can exist in a variety of oxidation states in aqueous solution.
- The +3 oxidation state is the most important for Ac, the later actinides Am to Md, and Lr.
- The +6 oxidation state is found in the actinyl ions AnO_2^{2+} for U, Np, Pu and Am.
- The +7 oxidation state is strongly oxidizing.
- Hydrolysis occurs readily.

3.1.3 Complex formation in aqueous solution

The f-orbitals of lanthanides are not involved in the bonding to ligands, which is electrostatic and non-directional. Crystal field effects, which are so important in transition metal chemistry, are absent in lanthanide chemistry, and the coordination chemistry of the lanthanides in aqueous solution is a battle against the formation of insoluble hydroxides. Lanthanide ions are considered as hard Lewis acids and generally form the most stable complexes with hard Lewis bases such as F^- or O-donors. The actinides exist in a larger variety of forms in aqueous solution, but like the Ln ions, all of these are hard Lewis acids, and show the greatest affinity for O and F donors. For a given ligand *e.g.* F^- the stability of the complexes generally increases in the order $AnO_2^+ < An^{3+} \leq AnO_2^{2+} < An^{4+}$.

Simple anions

Complex formation in aqueous solution requires first of all that coordinated H_2O molecules are displaced both from the metal ion and the incoming ligand. Because of the high enthalpies of hydration for Ln and An ions, complex formation is often an endothermic process and is only made possible by entropic effects.

The initial step in complex formation involves a loose interaction between the hydrated cation and the incoming ligand giving an outer-sphere complex. If a H_2O molecule is displaced from the inner coordination sphere of the metal ion by the incoming ligand, then an inner-sphere complex is formed. Some ligands are not able to displace coordinated H_2O from Ln ions and so form only outer-sphere complexes. The common anions Cl^-, Br^-, I^-, ClO_3^-, NO_3^-, sulfonate and trichloroacetate form predominantly outer sphere complexes with Ln^{3+} and actinide ions, whereas F^-, IO_3^-, SO_4^{2-} and OAc^- form predominantly inner-sphere complexes. It has been proposed that outer-sphere complexes are formed by ligands for which $pK_a < 2$ and inner sphere complexes are formed by ligands for which $pK_a > 2$.

- Complex formation in aqueous solution is often endothermic due to large ΔH°_{hyd} for Ln and An ions.
- Favourable entropy effects enable complex formation in aqueous solution.
- Outer sphere complexes formed with Cl^-, Br^-, I^-, ClO_3^-.
- Inner sphere complexes formed with F^-, IO_3^-, SO_4^{2-}, OAc^-.

Multidentate ligands

Simple anionic ligands do not form stable complexes with Ln^{3+} or actinide ions in aqueous solution: multidentate ligands are required. Lanthanide complexes are finding applications in medicine (*eg* contrast agents for magnetic resonance imaging, radiotherapeutic drugs, fluoroimmunoassay) which require stability under physiological conditions — aqueous solution at pH approximately 7 — and so preparation of complexes which are stable in aqueous solution is an important goal. Stability constants for a range of ligands are given in Table 3.2, and the structures of some of these are shown in Figure 3.5.

Aminopolycarboxylates

The aminopolycarboxylates are among the most important multidentate ligands and some of these (*e.g.* EDTA, DTPA) are available at very low cost. The stability constants for EDTA and DTPA complexes are plotted in Figure 3.6.

The EDTA complexes show a steady increase in log β as the radius of Ln^{3+} decreases, as expected for the steadily increasing electrostatic interaction. Log β for

TABLE 3.2 STABILITY CONSTANTS FOR LANTHANIDE COMPLEXES.

Ln	Cl^-	F^-	SO_4^{2-}	OAc^-	$P_3O_{10}^{5-}$	NTA^{3-} (LnL)	$EDTA^{4-}$	$DTPA^{5-}$	$DOTA^{4-}$	15-C-5[a]	18-C-6[a]	triglyme[b]	tetraglyme[b]
La	0.48	2.67	1.29	1.82	8	10.47	15.46	19.48		6.27	8.75	3.92	5.05
Ce	0.47	2.81	1.24	1.91	8.1	10.70	15.94	20.33					5.15
Pr	0.44	3.01	1.27	2.01	8.3	10.87	16.36	21.07		6.22	8.6	3.64	5.4
Nd	0.4	3.09	1.26	2.11	8.5	11.10	16.56	21.6				4.29	5.17
Sm	0.36	3.12	1.3	2.17	8.7	11.32	17.10	22.34		6.11	8.1	4.25	5.03
Eu	0.34	3.19	1.37	2.13	8.2	11.32	17.32	22.39	28.2		8.07		4.96
Gd	0.33	3.31	1.33	2.02	8.9	11.35	17.35	22.46					4.49
Tb	0.32	3.42	1.27	1.91	9.1	11.50	17.92	22.71	28.6	5.96	7.99	3.9	3.8
Dy	0.31	3.46	1.23	1.85	9.2	11.63	18.28	22.82		5.66	7.9	3.58	3.75
Ho	0.3	3.52	1.24	1.81	9.2	11.76	18.60	22.78					3.6
Er	0.26	3.54	1.23	1.79	9.5	11.90	18.83	22.74		5.33	7.67	3.7	3.73
Tm	0.25	3.56	1.15	1.83	9.7	12.07	19.30	22.72					3.63
Yb	0.24	3.58	1.15	1.84		12.20	19.48	22.62		5.53	7.5	4.56	3.7
Lu	0.23	3.61	1.09	1.85		12.32	19.80	22.44	29.2	5.83	7.2	4.41	3.72

Unless otherwise indicated, values are in 0.1 M ionic strength solution at 25°C and are taken from Martell, A.E. and Smith, R.M., Critical Stability Constants, Volumes 1–6, Plenum Press, New York, 1974–1989.

[a] Measured in propylene carbonate. Values from Bünzli and Pilloud (1989).

[b] Measured in propylene carbonate. Values from Barthélemy, Desreux and Massaux (1986).

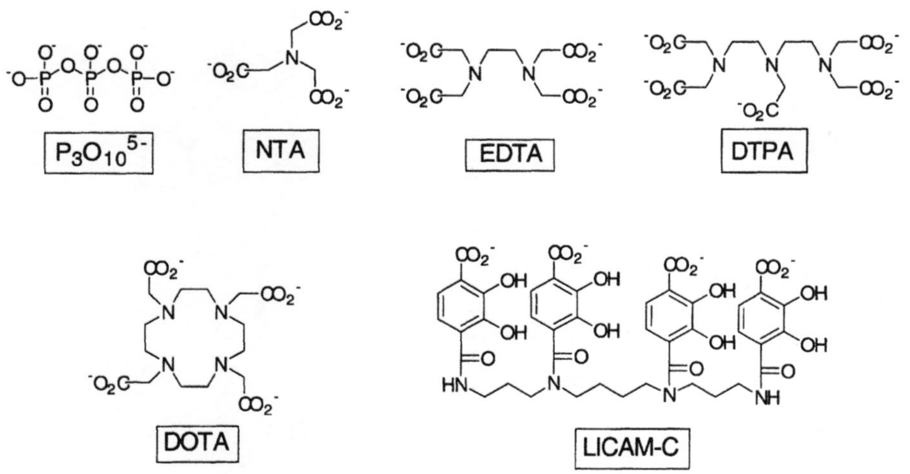

FIGURE 3.5 STRUCTURES OF SOME MULTIDENTATE LIGANDS REFERRED TO IN TABLE 3.2.

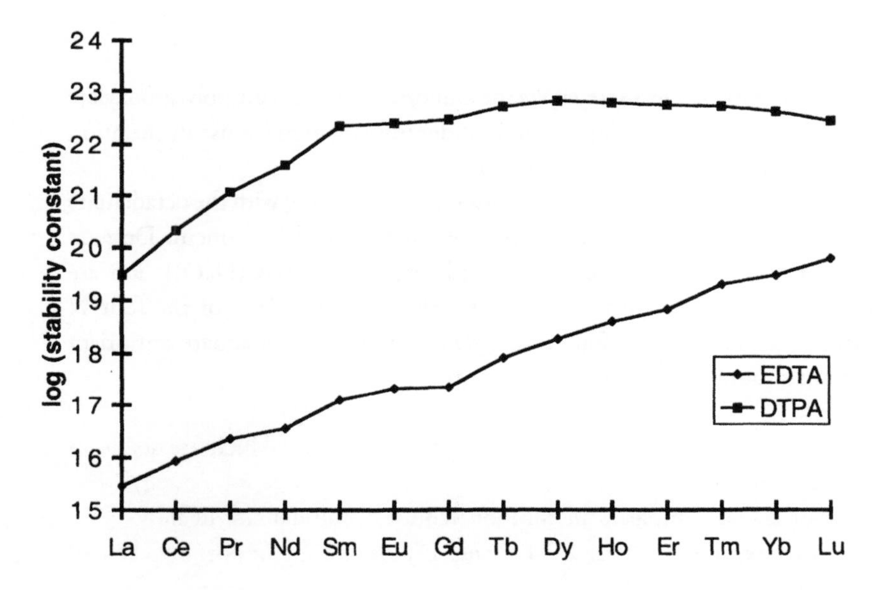

FIGURE 3.6 LOG β FOR FOR LN COMPLEXES WITH $EDTA^{4-}$ AND $DTPA^{5-}$.

DTPA goes through a maximum at Dy^{3+}, and this has been ascribed to decreased flexibility of DTPA compared with EDTA, which results in a decreased ability to coordinate to the smaller, later Ln^{3+} ions. The crystal structures of the 9-coordinate complexes $[Nd(EDTA)(H_2O)_3]^-$ and $[Nd(DTPA)(H_2O)]^{2-}$ are shown in Figure 3.7.

Actinide ions also form stable complexes with EDTA and DTPA. The DTPA complexes are of particular interest as they are used in chelation therapy to treat cases of exposure to An^{4+} or An^{3+}. Log β for $[Pu(DTPA)]^-$ is 29.9; for $[Am(DTPA)]^{2-}$

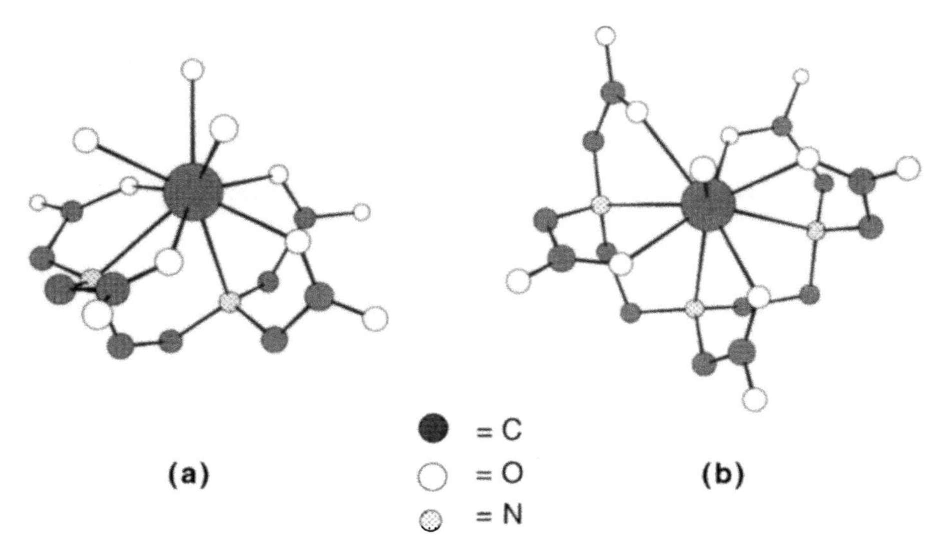

FIGURE 3.7 STRUCTURES OF LANTHANIDE AMINOPOLYCARBOXYLATES. (A) $[Nd(EDTA)(H_2O)_3]^-$ (B) $[Nd(DTPA)(H_2O)]^{2-}$.

it is 22.9, and the complexes are excreted *via* the kidneys. A Pu^{4+} specific polycarboxylate ligand, LICAM-C, has been developed and is under investigation for use in chelation therapy. Its structure is shown in Figure 3.5.

The most stable complexes of Ln^{3+} in aqueous solution are those with the octadentate DOTA ligand, for which stability constants were determined by Loncin, Desreux and Merciny (1986). Two views of the 9-coordinate $[Gd(DOTA)(H_2O)]^-$ ion are shown in Figure 3.8. The large Gd^{3+} ion sits well above the plane of the four N atoms which, together with the four carboxylate groups, form a square antiprism of donor atoms around the metal.

- Applications in medicine and physiology require complexes which are stable in aqueous solution.
- Formation of stable complexes in solution requires multidentate ligands.
- Aminopolycarboxylates such as EDTA and DTPA are an important class of ligands for both Ln and An.

3.2 COMPLEXES WITH NEUTRAL O-DONOR LIGANDS

All of the complexes described in this section and the remainder of Chapter 3 must be prepared in non-aqueous solution, although not necessarily under rigorously anhydrous conditions.

3.2.1 Phosphine, arsine and amine oxides

Complexes of lanthanides and actinides with amine oxides such as pyridine N-oxide are well known, and the structure of the $[La(C_5H_5NO)_8]^{3+}$ ion, which has square antiprismatic coordination geometry, is shown in Figure 3.9.

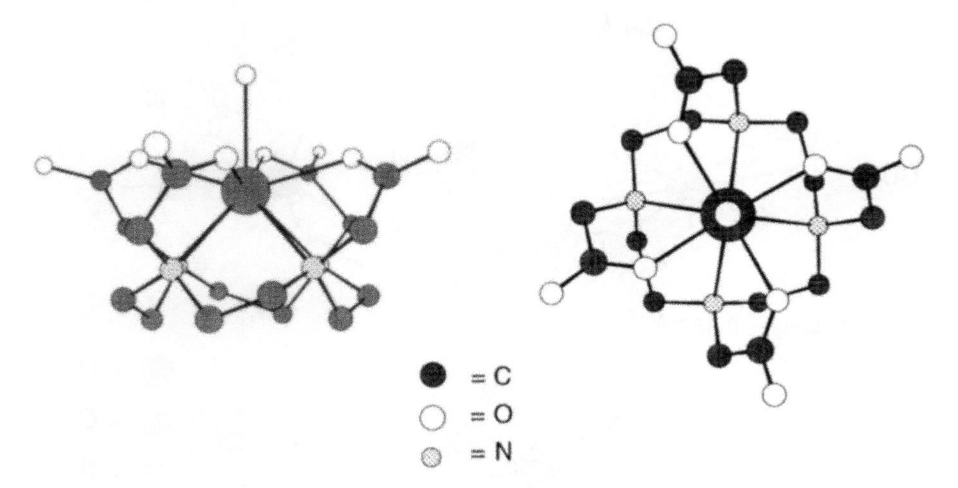

FIGURE 3.8 STRUCTURE OF $[Gd(DOTA)]^{-}$.

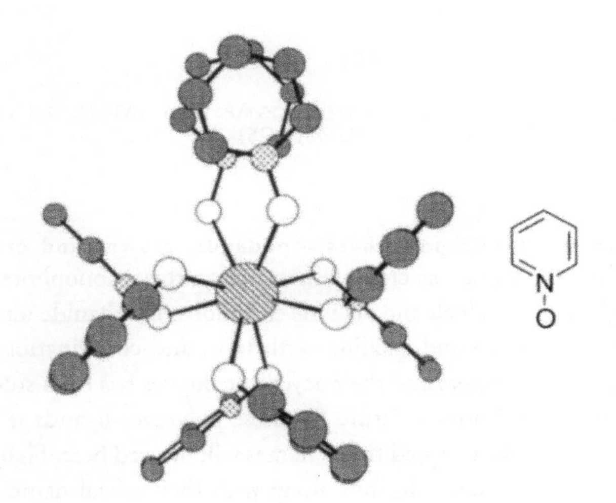

FIGURE 3.9 STRUCTURE OF $[La(pyridine\text{-}N\text{-}oxide)_8]^{3+}$.

However, phosphine oxides and arsine oxides bind much more strongly to Ln and An ions. Triphenylphosphine oxide, $Ph_3P{=}O$, forms adducts with many Lewis acidic lanthanide complexes in non-aqueous solution and is often used to stabilize complexes which would otherwise be very coordinatively unsaturated. The related ligand HMPA, $(Me_2N)_3P{=}O$, is also a very strong donor and is used to increase the activity of SmI_2 (see Chapter 5) by formation of the 6-coordinate complex $[SmI_2(HMPA)_4]$. The P=O donor group is important in nuclear fuel processing where a complex of $(BuO)_3P{=}O$ with $UO_2(NO_3)_2$ is used in the solvent extraction purification of uranium as described in Section 3.6. The structure of $[UO_2(NO_3)_2\{(MeO)_3P{=}O\}]$ is shown in Figure 3.10.

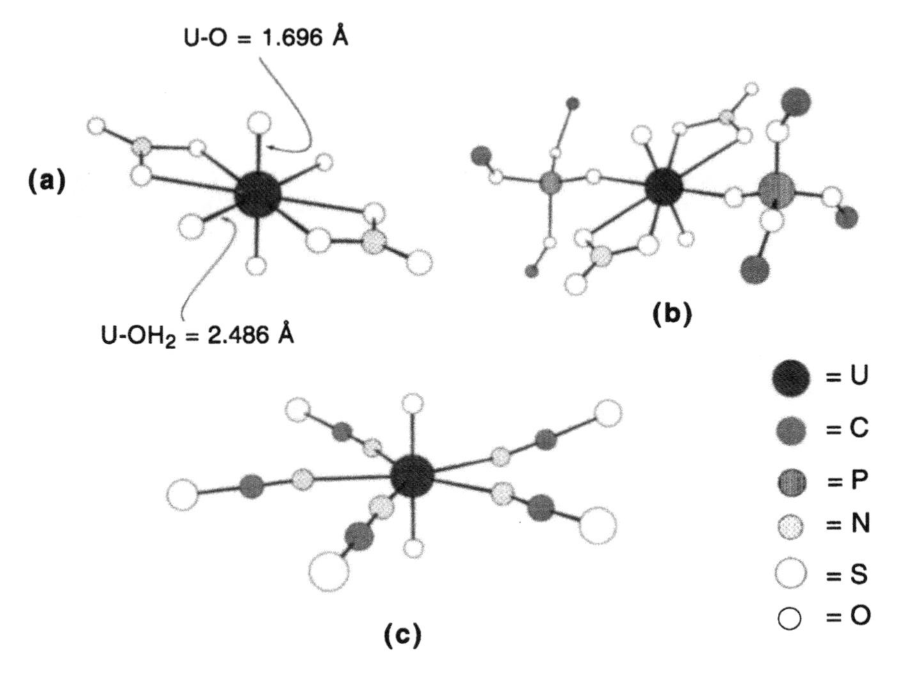

FIGURE 3.10 STRUCTURES OF URANYL COMPLEXES (A) $[UO_2(NO_3)_2(H_2O)_2]$; (B) $[UO_2(NO_3)_2\{(MeO)_3P{=}O\}]$; (C) $[UO_2(NCS)_5]^{3-}$.

3.2.2 Acyclic and cyclic polyethers – podands, crowns and cryptands

The cyclic polyethers known as crown ethers are synthetic ionophores which show selective complexation of alkali and alkaline earth ions. Lanthanide ions have similar radii to those of the alkali and alkaline earth ions, and coordination chemistry of lanthanides with crown ethers and their acyclic analogues has been studied since the late 1970's. The IUPAC nomenclature for these polyether ligands is very cumbersome and so the generally accepted trivial names will be used here. Figure 3.11 shows the structures of representative ligands along with their trivial names. The coordination chemistry of lanthanides with these ligand systems has been reviewed by Bünzli and Wessner (1984). All of the complexes described in this Section must be prepared in non-aqueous solvents as these ligands are not able to displace coordinated H_2O molecules of the Ln^{3+} aquo ions in aqueous solution. Complexes have been characterized with a range of counterions including Cl^-, NCS^-, ClO_4^- and NO_3^-. Because of its good coordinating properties NO_3^- is probably the most used counterion in structural studies.

Ln^{3+} form a range of complexes with the crown ethers. These are simple 1:1 complexes as well as structures with 2:1 and 3:4 metal:ligand ratios. 18-C-6 has a cavity size which is well-matched to the ionic radius of the larger Ln^{3+} ions and the crystal structure of $[La(NO_3)_3(18\text{-}C\text{-}6)]$ shows that the La^{3+} ion is coplanar with the six O-donors of the crown ether. The flexibility of 18-C-6 allows it to pucker and bind reasonably effectively to the smaller later lanthanides. Compounds with a metal:ligand ratio of 4:3 are often obtained with lanthanide nitrates where the crown ether

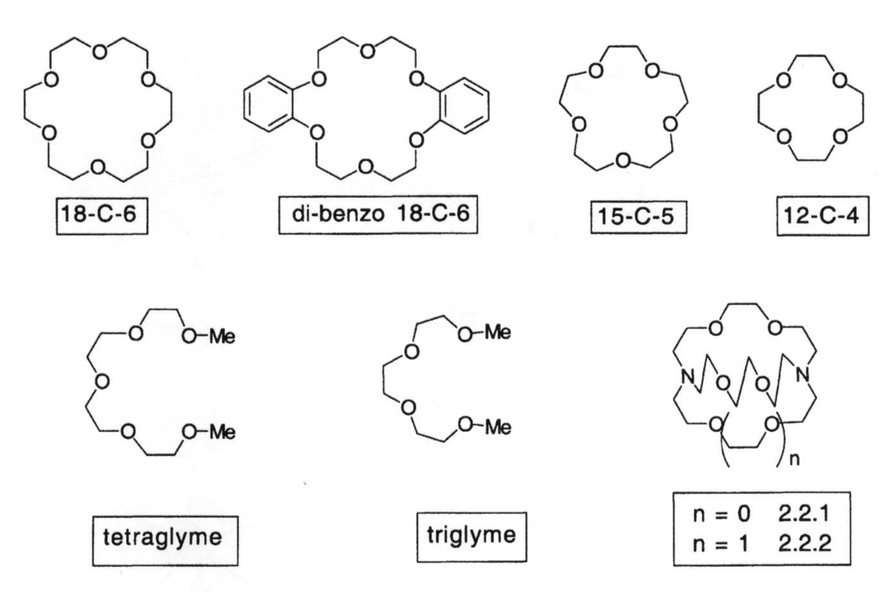

FIGURE 3.11 STRUCTURES OF CROWN, PODAND AND CRYPTAND LIGANDS.

complex is $[Ln(NO_3)_2(18\text{-}C\text{-}6)]^+$ and the 12-coordinate $[Ln(NO_3)_6]^{3-}$ is present as a counterion. Dibenzo-18-C-6 is much less flexible than the unsubstituted ligand and in the presence of the relatively strongly coordinating bidentate NO_3^- counterion, simple 1:1 complexes can only be isolated for the large early Ln^{3+} ions La to Nd.

1:1 complexes of Ln^{3+} with 15-C-5 or 12-C-4 are known for La-Lu with NO_3^- counterions. The large Ln^{3+} ions cannot fit within the cavity of these ligands, and crystal structures of the complexes show the Ln^{3+} ion sitting above the plane of the macrocycle. Structures of complexes with 15-C-5 and 12-C-4 are shown in Figure 3.12.

The acyclic analogues of 15-C-5 and 12-C-4 are tetraglyme and triglyme. These ligands form well characterized complexes with all the lanthanides. Unlike their crown ether analogues, tetraglyme and triglyme wrap around the lanthanide ion which sits within the plane of the four or five O-donors. The structure of a La complex with tetraglyme is shown in Figure 3.13.

Stability constants have been determined in anhydrous propylene carbonate solvent for Ln^{3+} complexes with 18-C-6, 15-C-5 by Bünzli and Pilloud (1989) and the acyclic polyethers tetraglyme and triglyme by Barthélemy, Desreux and Massaux (1986). These are plotted in Figure 3.14. It is found that there is a good correlation between stability of crown ether complexes and the match of ionic radius to cavity size. Thus the 18-C-6 complexes show the highest stability. Complexes with the large 21-C-7 macrocycle show lower stability than those with 15-C-5. As a result of the macrocyclic effect, complexes with tetraglyme have lower stability than those with the cyclic analogue 15-C-5. The stability of tetraglyme complexes falls off quite dramatically for the late lanthanides Tb – Lu. This observation has been explained by steric repulsions between the terminal Me groups of tetraglyme as the ligand wraps itself around the smaller Ln^{3+} ions.

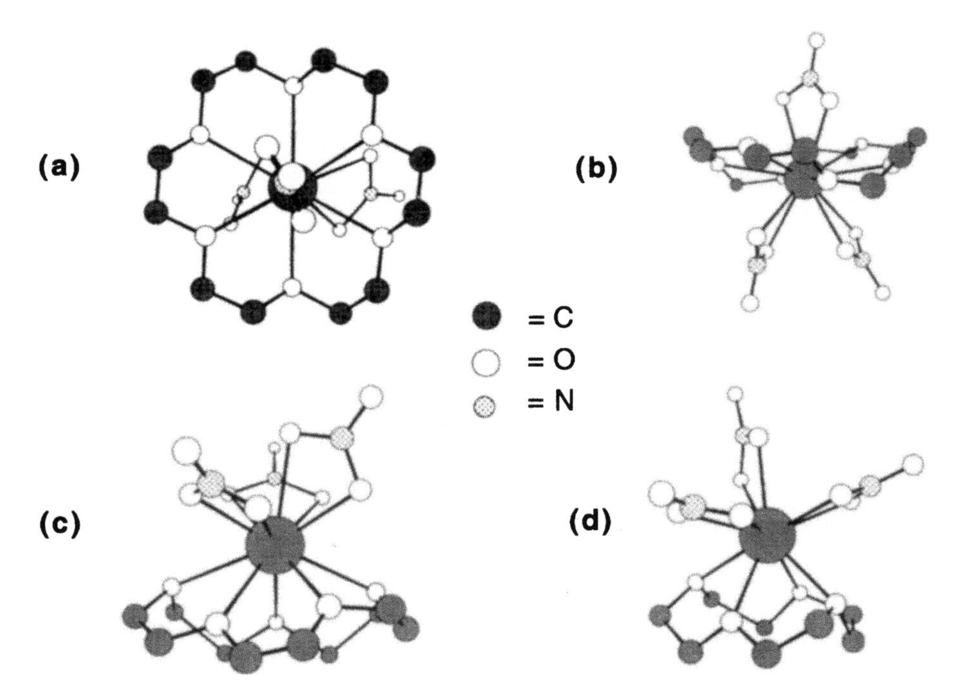

FIGURE 3.12 STRUCTURES OF LN COMPLEXES WITH CROWN ETHERS. (A) [$LA(NO_3)_3$(18-C-6)] VIEWED PERPENDICULAR TO PLANE OF CROWN ETHER (B) [$LA(NO_3)_3$(18-C-6)] VIEWED PARALLEL TO EQUATORIAL PLANE (C) [$EU(NO_3)_3$(15-C-5)] (D) [$EU(NO_3)_3$(12-C-4)].

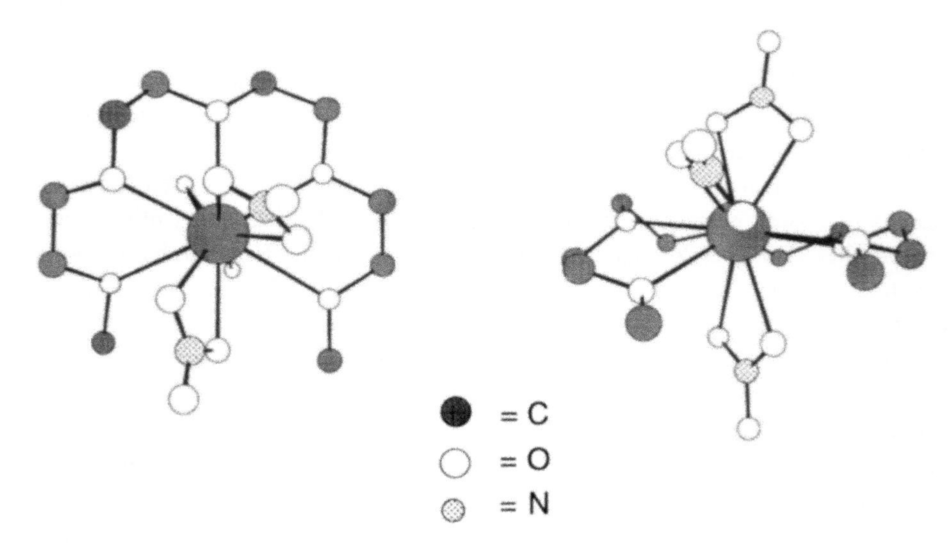

FIGURE 3.13 TWO VIEWS OF THE STRUCTURE OF [$LA(NO_3)_3$(TETRAGLYME)].

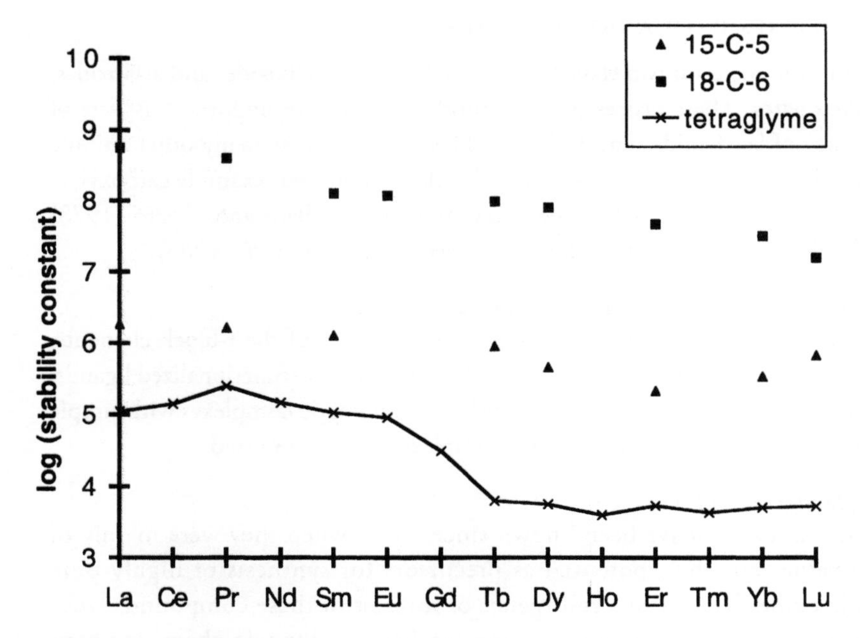

FIGURE 3.14 LOG β FOR COMPLEXES OF LN^{3+} WITH 18-C-6, 15-C-5 AND TETRAGLYME (MEASURED IN PROPYLENE CARBONATE).

The macrobicyclic cryptand ligands (2.1.1), (2.2.1) and (2.2.2) (see Figure 2.4) form complexes with all the lanthanides. Although they are kinetically inert towards dissociation in aqueous solution (unlike complexes of crowns and podands) they do not have a sufficiently high formation constant to allow their preparation in H_2O. Instead they must be prepared under quite strictly anhydrous conditions in weakly coordinating solvents such as acetonitrile. Once made these complexes are very stable and are of interest for applications in aqueous solution. Their stability constants in aqueous solution have been measured by Burns and Baes (1981). The stabilities of Ln^{3+} cryptates are found to be similar to those of the alkaline earths, and lower than those of divalent transition and post-transition metal ions. They show very little dependence on ionic radius. Eu^{2+} cryptates are found to be much more stable than their Eu^{3+} analogues.

Complexes of the actinides with crown ethers are also known, especially for the UO_2^{2+} ion. However they have not been subjected to such extensive study as their lanthanide counterparts.

- Complexes of Ln and An with crown ethers must be prepared in non-aqueous solution.
- The stability of crown ether complexes correlates well with the match of cavity size to ionic radius.
- Cryptate complexes of Ln are kinetically inert in aqueous solution.

3.3 COMPLEXES WITH ANIONIC O-DONORS

This Section will focus on two classes of anionic O-donors: alkoxides and aryloxides, and β-diketonates. These classes of compound will illustrate important aspects of the chemistry of lanthanides and actinides. There are of course many other anionic O-donors which form complexes with the f-block elements. For example carboxylate complexes with the actinides have been reviewed by Casellato and Vigato (1978) and lanthanide carboxylates have been reviewed by Ouchi *et al.* (1988).

3.3.1 Alkoxide and Aryloxide Complexes

There are numerous heteroleptic alkoxides and aryloxides of the f-block elements, as well as heterometallic complexes and complexes with donor-functionalized ligands. This section will be limited almost exclusively to homoleptic complexes with simple monodentate ligands. Similarly, siloxide complexes will be omitted.

Lanthanide(III) complexes

Lanthanide alkoxides have been known since 1958, when they were mainly of curiosity value, but their potential as precursors for synthesis of highly pure lanthanide oxides has led to a resurgence of interest in these compounds since the early 1980's. Major reviews covering lanthanide alkoxide chemistry have been written by Mehrotra, Singh and Tripathi (1991) and by Mehrotra and Singh (1997). Anwander (1996) has reviewed routes to monomeric lanthanide alkoxides.

The large size of the Ln^{3+} ions results in lanthanide alkoxides adopting a variety of structures dictated by the demands of the alkoxide ligand. $[Ln(OR)_3]$ where R = Me, Et or Ph are involatile and virtually insoluble in benzene; they have polymeric structures where the lanthanide coordination number is increased by formation of Ln-OR-Ln bridges. With the more sterically demanding OPr^i ligand, molecular weight determinations show that a tetrameric species $[Ln(OPr^i)_3]_4$ exists in solution although crystals isolated from solution have the formula $[Ln_5O(OPr^i)_{13}]$ where the central O atom is derived from C-O cleavage in a OPr^i ligand as shown in Scheme 3.1. Addition of $Al(OPr^i)_3$ gives a heterometallic compound where the Ln achieves coordinative saturation by forming Ln-OPr^i-Al bridges. A trimeric complex $[Ln_3(OBu^t)_4(\mu_2\text{-}OBu^t)_3(\mu_3\text{-}OBu^t)_2(Bu^tOH)_2]$ is obtained for the more sterically demanding OBu^t ligand.

Increasing the steric bulk of the ligand leads to decreasing degrees of oligomerization: O-(2,6-Me_2-C_6H_3) gives monomeric complexes in the presence of excess THF, and a dimeric complex when crystallized from toluene as shown in Figure 3.15. With the very bulky O-(2,6-Bu^t_2-C_6H_3) ligand a monomeric complex is formed. This complex has been characterized by X-ray diffraction and its structure is shown in Figure 3.15. The structure is very similar to that adopted by other three-coordinate f-block complexes such as the tris(silylamides) and the tris(bis(trimethylsilyl)methyl) complexes. Adduct formation with small Lewis bases can occur and the five coordinate $[Ce(O\text{-}2,6\text{-}Me_2C_6H_3)_3(RCN)_2]$ has been reported.

$[Nd(O\text{-}2,6\text{-}Pr^i_2C_6H_3)_3]$ has a novel means of achieving coordinative saturation: an unusual η^6 π interaction with an aryl ring has been observed both in solution

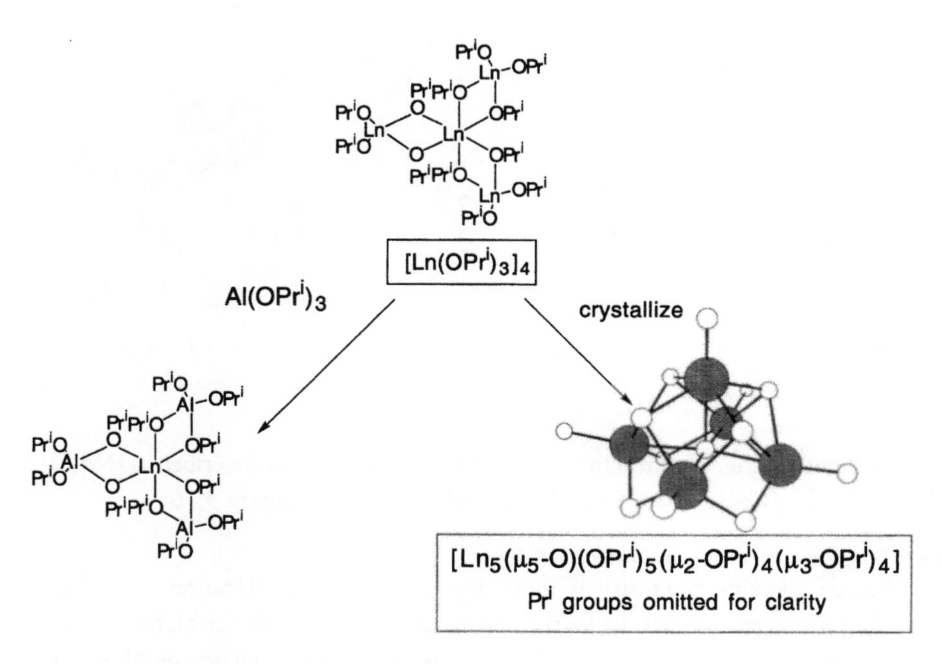

SCHEME 3.1 REACTIONS OF LN(OPR^i)$_3$.

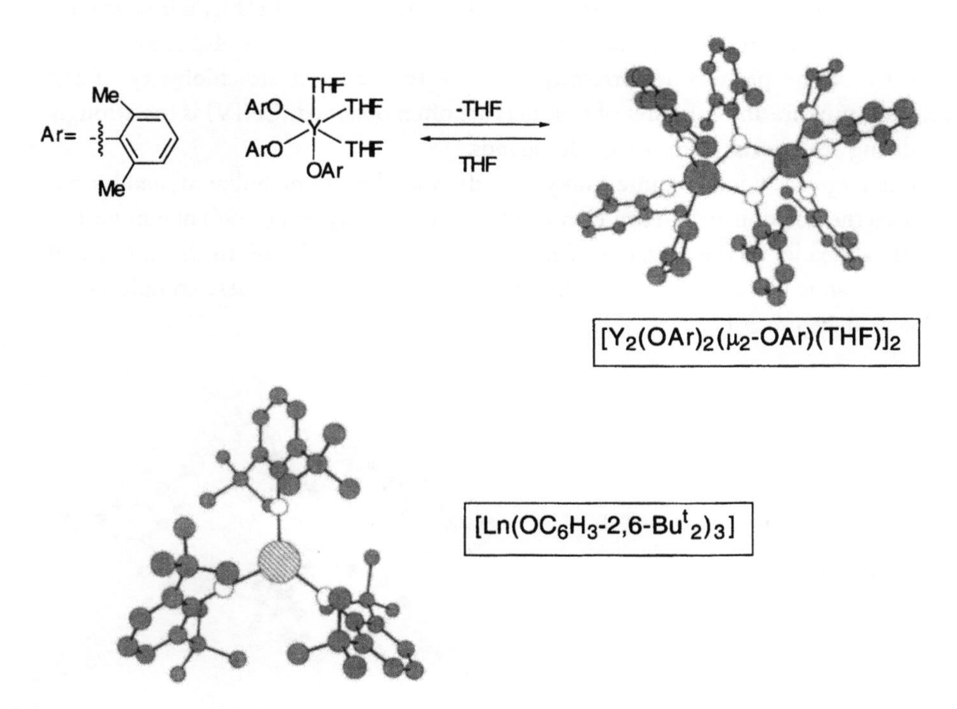

FIGURE 3.15 STRUCTURES OF LANTHANIDE ARYLOXIDES.

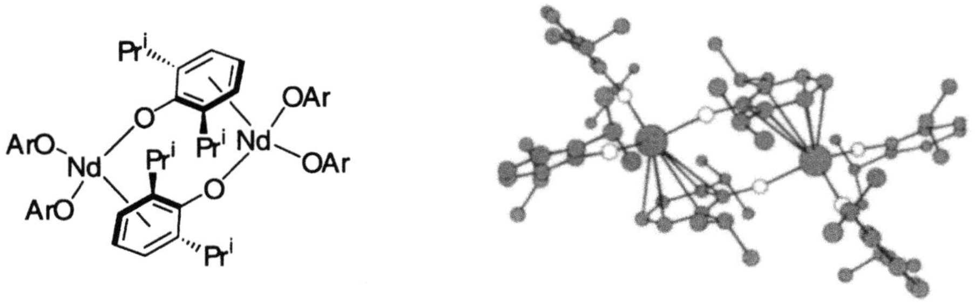

FIGURE 3.16 STRUCTURE OF $[Nd(O\text{-}2,6\text{-}Pr^i{}_2C_6H_3)_2(\eta^6\text{-}\mu_2\text{-}O\text{-}2,6\text{-}Pr^i{}_2C_6H_3)]_2$.

and in the solid state. The structure of the resulting dimer is best described as $[Nd(O\text{-}2,6\text{-}Pr^i{}_2C_6H_3)_2(\eta^6\text{-}\mu_2\text{-}O\text{-}2,6\text{-}Pr^i{}_2C_6H_3)]_2$ and is shown in Figure 3.16.

Ln alkoxides in other oxidation states

Lanthanide alkoxide chemistry is inevitably dominated by oxidation state (III). However, it is expected that the hard alkoxide ligand should stabilize high oxidation states, and complexes of Ce(IV) with OBu^t are well known. These complexes are synthesized by the metathesis reaction of $[NH_4]_2[Ce(NO_3)_6]$ with $NaOBu^t$. A range of neutral complexes of general formula $[Ce(OBu^t)_n(NO_3)_{4-n}(THF)_x]$ is known with n = 1 to 4, and anionic 'ate' complexes incorporating Na ions are also formed. The identity of the product is extremely sensitive to the exact stoichiometry of the reaction mixture and mixtures of products are often obtained. Ce(IV) is too strongly oxidizing to co-exist with aryloxide ligands.

The large Ln^{2+} ions require bulky ligands, and the use of bulky aryloxides has allowed the isolation by Van den Hende, Hitchcock and Lappert (1994) of monomeric Yb(II) alkoxides in the presence of neutral donors such as THF. In the absence of donor ligands the complexes exist as dimers. The structures of these complexes are shown in Figure 3.17.

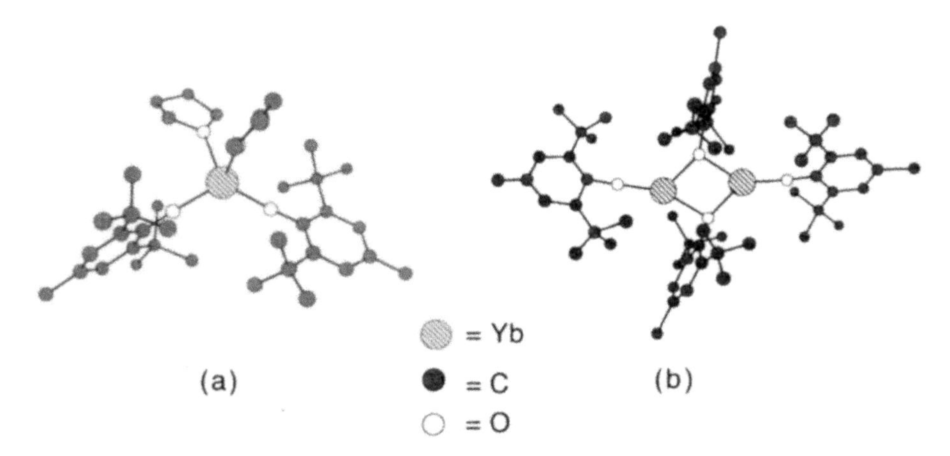

FIGURE 3.17 STRUCTURES OF YB(II) ARYLOXIDES. (a) $[Yb(OC_6H_2\text{-}2,6\text{-}Bu^t{}_2\text{-}4\text{-}Me)_2(THF)_2]$. (b) $[Yb(\mu_2\text{-}OC_6H_2\text{-}2,6\text{-}Bu^t{}_2\text{-}4\text{-}Me)(OC_6H_2\text{-}2,6\text{-}Bu^t{}_2\text{-}4\text{-}Me)]_2$.

Reactivity

The reactivity of lanthanide alkoxides is dominated by their Brønsted basicity: they react readily with protic reagents such as H_2O, β-diketones, silanols and organic acids. This reactivity has been exploited in their use as catalysts for the anionic ring-opening polymerization of lactones described in Chapter 5. Although they generally achieve coordinative saturation by the formation of oligomers *via* OR bridging they can also form adducts with Lewis bases such as THF, pyridine and diglyme ($MeO(CH_2CH_2O)_2Me$) to form monomeric species. This combination of Lewis acidity and Brønsted basicity has led to the use of lanthanide alkoxides as catalysts for organic synthesis. Heterometallic species are formed by addition of other metal alkoxides, for example the addition of $Al(OPr^i)_3$ to $Ln(OPr^i)_3$, as shown in Scheme 3.1.

Applications

The huge increase in interest in lanthanide alkoxides since the early 1980's has been driven by their potential as precursors to high purity oxides which has been reviewed by Hubert-Pfalzgraf (1995). MOCVD (Metal Organic Chemical Vapour Deposition) involves thermal decomposition of metal alkoxide to metal oxide and requires volatile precursors. Volatility can be enhanced by the use of bulky (*e.g.* O-2,6-$Bu^tC_6H_3$) or fluorinated ligands. For example $[Ln(O\text{-}2,6\text{-}Bu^tC_6H_3)_3]$ sublime at approximately 200°C at 10^{-3} mm Hg, and $[Ln(OCMe(CF_3)_2)_3(diglyme)]$ sublime at approximately 150°C and 10^{-2} mmHg.

The sol-gel process involves controlled hydrolysis of metal alkoxide to produce metal hydroxide and ultimately metal oxide with elimination of alcohol.

$$2Ln(OR)_3 + 6H_2O \rightarrow 2Ln(OH)_3 + 6ROH \rightarrow Ln_2O_3 + 3H_2O$$

The hydrolysis of molecular heterometallic alkoxides allows formation of mixed metal oxides of predetermined stoichiometry. This area has been reviewed by Chandler, Roger and Hampden-Smith (1993). The use of lanthanide alkoxides in catalysis is discussed in Chapter 5.

Actinides

Actinide alkoxides have more varied chemistry than their lanthanide counterparts because of the range of oxidation states available; their chemistry has been reviewed by Van der Sluys and Sattelberger (1990). Not surprisingly, most of the work on actinide alkoxides has been carried out with thorium and uranium. A small number of alkoxides of protactinium, neptunium and plutonium have also been prepared and characterized but will not be discussed further here. Investigation of actinide alkoxides began in 1954 with the report of $[Th(OPr^i)_4]$, prepared in high yield by the metathesis of $ThCl_4$ with $NaOPr^i$ in refluxing Pr^iOH. Molecular weight measurements showed this complex to exist as a tetramer $[Th(OPr^i)_4]_4$ in benzene solution or as a dimer in Pr^iOH. A range of Th(IV) alkoxides and aryloxides, both homeleptic and heteroleptic, has now been synthesized.

Uranium shows the greatest range of alkoxide and aryloxide complexes, in oxidation states ranging from (III) to (VI). The first uranium alkoxide was $[U(OPr^i)_4]$, prepared in 1956 by a metathesis reaction with UCl_4 similar to that used for the analogous Th(IV) complex. Since then a wide range of U(IV) alkoxides and aryloxides have been synthesized by this route. The best route to uranium(III) alkoxides is the alcoholysis of $[U\{N(SiMe_3)_2\}_3]$: metathesis reactions of *in situ* prepared $UCl_3(THF)_n$ do not give isolable products. Uranium(V) alkoxides can be prepared either by metathesis reactions of UCl_5 with alkali metal alkoxides or by oxidation of uranium(IV) alkoxides *e.g.* with Br_2 as shown in Scheme 3.2. $[U(OR)_5]$ synthesized in this way can then be used as a starting material for other alkoxides *via* alcohol exchange. Highly oxidizing uranium(VI) alkoxides can be synthesized by oxidation of U(V) complexes as shown in Scheme 3.2 or by metathesis reactions of UF_6.

$$[U(OEt)_4] + 0.5\ Br_2 \xrightarrow{EtOH} \text{"}U(OEt)_4Br\text{"} \xrightarrow{EtOH/NaOEt} [U(OEt)_5] + NaBr$$

$$2NaU(OEt)_6 + (PhCO)_2O_2 \xrightarrow{EtOH} 2[U(OEt)_6] + 2NaO_2CPh$$

SCHEME 3.2 PREPARATION OF URANIUM ALKOXIDES.

The structures of actinide alkoxides are, like those of lanthanide alkoxides, determined by steric factors, and dimerization or oligomerization occurs in order to achieve coordinative saturation of the An ion. This is generally *via* the formation of μ_2-OR bridges, but $[U(O\text{-}2,6\text{-}Pr^i{}_2C_6H_3)_2]_2$, like its Nd analogue described earlier, has η^6 π-aryl bridges. The structures of a four-coordinate U(IV) aryloxide and a six-coordinate U(V) alkoxide are shown in Figure 3.18.

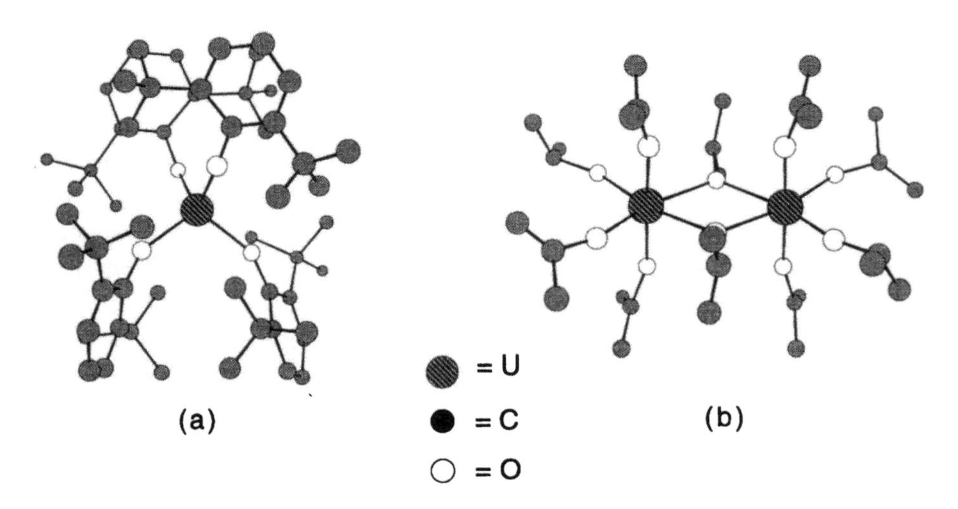

FIGURE 3.18 STRUCTURES OF URANIUM ALKOXIDES AND ARYLOXIDES (A) $[U(OC_6H_3\text{-}2,6\text{-}Bu^t{}_2)_4]$ (B) $[U_2(OPr^i)_{10}]$.

Actinide alkoxides, like their lanthanide analogues, are all extremely moisture-sensitive and must be handled under rigorously anhydrous conditions. Adduct formation with Lewis bases is also a common reaction, and even phosphines will form adducts of the type $[An(OR)_4(dmpe)]$ (dmpe = $Me_2P(CH_2CH_2)PMe_2$). One motivation for the investigation of U alkoxide chemistry has been the potential use of volatile U alkoxides such as $[U(OMe)_6]$ in isotope separation or enrichment as described by Miller *et al.* (1979).

- The steric bulk of ^-OR determines the structures of Ln and An alkoxides.
- $Ln(OMe)_3$ are polymeric; $Ln(OAr)_3$ (Ar = $^-OC_6H_3$-2,6-Bu^t_2) are monomeric.
- Ln and An alkoxides are Brønsted bases. Many of them are also Lewis acids.
- $Ln(OR)_3$ are precursors to high purity Ln oxides *via* MOCVD or sol-gel routes.
- Alkoxides of U are known in oxidation states +2 to +6.

$$LnCl_3{\cdot}3Pr^iOH + 3NaOPr^i \xrightarrow{Pr^iOH} Ln(OPr^i)_3 + 3Pr^iOH + 3NaCl$$

$$ThCl_4 + 4NaOPr^i \xrightarrow{Pr^iOH} Th(OPr^i)_4 + 4\,NaCl$$

$$Ln + 3Pr^iOH \xrightarrow{HgCl_2\ \text{catalyst}} Ln(OPr^i)_3 + 3/2H_2$$

$$Ln(OPr^i)_3 + 3ROH \longrightarrow Ln(OR)_3 + 3Pr^iOH$$

$$U\{N(SiMe_3)_2\}_3 + 3ROH \longrightarrow U(OR)_3 + 3HN(SiMe_3)_2$$

SCHEME 3.3 SYNTHETIC ROUTES TO LANTHANIDE AND ACTINIDE ALKOXIDES.

Synthesis

Some general preparative routes applicable to both lanthanide and actinide alkoxides are summarized in Scheme 3.3. The metathesis route using alkali metal alkoxides and Ln or An halides is probably the most widely used, but it sometimes has the disadvantage of giving products containing alkali metal or halide. Because the metals are very electropositive, direct reaction with the appropriate alcohol is sometimes possible. This method avoids contamination of product, but is not universally applicable and may require a catalyst such as $HgCl_2$. Metal atom synthesis has been applied to lanthanide alkoxides but requires specialist apparatus. The 'silylamide route' — reaction of a metal silylamide complex with a protic reagent — will be described in more detail in Section 3.4, and has been applied successfully to alkoxide synthesis. Although synthesis of the starting material is required, this method has the advantages of producing halide- and alkali-metal free product and of giving volatile and easily removed amine as the only by-product.

3.3.2 β-Diketonate Complexes

Lanthanides

Lanthanide β-diketonates [Ln(diket)$_3$] have been known since the early years of the 20th century; systematic investigations of their chemistry began in the late 1940's, and because of the applications of these complexes, particularly in NMR spectroscopy (See Chapter 2) and materials synthesis, they still attract much current interest. The complexes display varying structures depending on the steric demands of the β-diketonate and the ionic radius of the Ln ion as shown in Figure 3.19.

For example [Ln(acac)$_3$] adopt polymeric structures in the solid state whereas complexes with the more sterically demanding thd ligand are dimeric for large Ln and monomeric for small Ln. Anionic tetrakis(β-diketonate) complexes are known for the less sterically demanding ligands such as acac, dbm and hfa. A summary of the most common β-diketonates is given in Table 3.3.

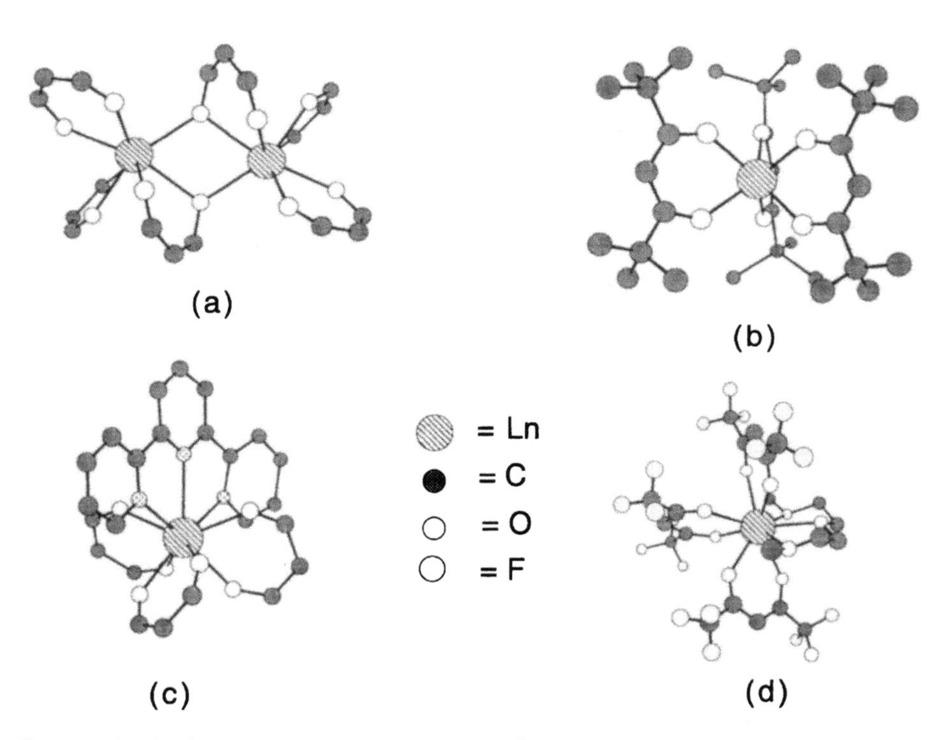

FIGURE 3.19 STRUCTURES OF LANTHANIDE β-DIKETONATES (A) [PR(THD)$_3$]$_2$ (Bu^t GROUPS OMITTED FOR CLARITY) (B) [LU(THD)$_3$] (C) [LA(ACAC)$_3$(TERPY)] (Me GROUPS OMITTED FOR CLARITY) (D) [LA(HFA)$_3$(DIGLYME)].

TABLE 3.3 β-DIKETONATE LIGANDS $(R^1C(O)CHC(O)R^2)^-$.

abbreviation	R^1	R^2
acac	Me	Me
dbm	Ph	Ph
fod	$CF_3CF_2CF_2$	Bu^t
hfa	CF_3	CF_3
thd	Bu^t	Bu^t
tta	C_4H_3S	CF_3

Apart from polymeric $[Ln(acac)_3]$ all the lanthanide β-diketonates are soluble in organic solvents, and several are volatile. The volatility and thermal stability of $[Ln(thd)_3]$ made possible the separation of Ln by GC of benzene solutions of $[Ln(thd)_3]$ reported by Eisenstraut and Sievers (1965).

Reactivity of $[Ln(diket)_3]$ is really limited to adduct formation with Lewis bases; isolable adducts with strong donors such as pyridine or Ph_3PO have been characterized by X-ray diffraction. The number of donor ligands is determined by the steric requirements of diket and of L; although they are not isolable, labile adducts with alcohols and ethers are observed in solution. Apart from formation of H_2O adducts they are relatively unreactive with water and may be handled in air without decomposition.

Adduct formation is exploited in the application of these complexes as NMR shift reagents (see Chapter 2) and in Lewis acid catalysis (see Chapter 5). GC separation of $[Ln(diket)_3]$ is not of current interest, but volatile complexes are now of great interest as precursors for materials synthesis by MOCVD. Enhanced volatility is achieved with fluorinated ligands such as hfa, and formation of adducts with polyethers such as diglyme $(MeO(CH_2CH_2O)_2Me)$ reported by Malandrino *et al.* (1995) produces stable monomeric complexes which sublime intact. For example $[La(hfa)_3(diglyme)]$ sublimes at 65–70°C and 10^{-3} mm Hg.

The photophysics of lanthanide β-diketonates has been well-studied: sensitized emission from Eu complexes under uv irradiation was first observed in 1942. Tris(β-diketonates) of Eu and Tb are being investigated as red and green emitters in electroluminescent devices (See Chapter 2). In addition to all these practical applications, one lanthanide β-diketonate complex shows the intriguing property of triboluminescence. When crystals of $[HNEt_3][Eu(dbm)_4]$ are crushed they emit bright orange-red flashes characteristic of Eu(III) emission, but only if they have been crystallized from the correct solvent: disordered crystals from MeOH are triboluminescent whereas crystals of the CH_2Cl_2 solvate are not. This phenomenon has been investigated by Sweeting and Rheingold (1987).

Actinides

Many actinide complexes are known with β-diketonate ligands. These include the 8-coordinate $[An(acac)_4]$ complexes, which have high stability constants and can be

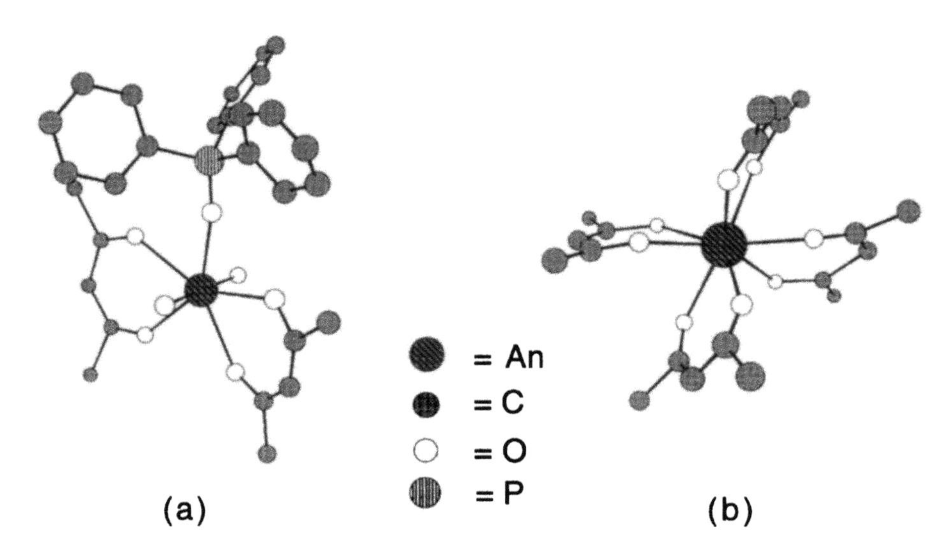

FIGURE 3.20 STRUCTURES OF ACTINIDE β-DIKETONATES. (a) $[UO_2(acac)_2(Ph_3PO)]$ (b) $[Th(acac)_4]$.

used to extract An^{4+} ions from aqueous into organic solution. These complexes crystallize with either dodecahedral or square antiprismatic coordination around An. Use of the tta ligand instead of acac gives complexes with enhanced solubility in organic solvents. Actinyl ions also form complexes with β-diketonates; in these complexes the β-diketonate ligands occupy the equatorial plane, and coordination of a further neutral ligand may be possible. The structures of two actinide β-diketonates are shown in Figure 3.20.

Synthesis

The most important route to lanthanide and actinide β-diketonates is metathesis of a metal salt such as halide or nitrate with sodium or ammonium β-diketonate. Reaction of an acid sensitive precursor such as $[Ln(OR)_3]$ with β-diketone has been used to prepare anhydrous $[Ln(acac)_3]$, and reaction of Hhfa with Ln_2O_3 and diglyme in hexane gives $[Ln(hfa)_3(diglyme)]$.

- Steric factors determine the structures of β-diketonate complexes: $[Lu(thd)_3]$ is a six-coordinate monomer, $[La(thd)_3]_2$ is a seven-coordinate dimer.
- $[Ln(diket)_3$ are Lewis acids. This is exploited in use as NMR shift reagents (Chapter 2) and catalysts (Chapter 5).
- Ln and An β-diketonates are relatively unreactive with H_2O, except for adduct formation.
- Volatile $[Ln(diket)_3]$ are precursors to Ln oxides *via* MOCVD.
- β-diketonates have been used in extraction of An ions from aqueous to organic solution.

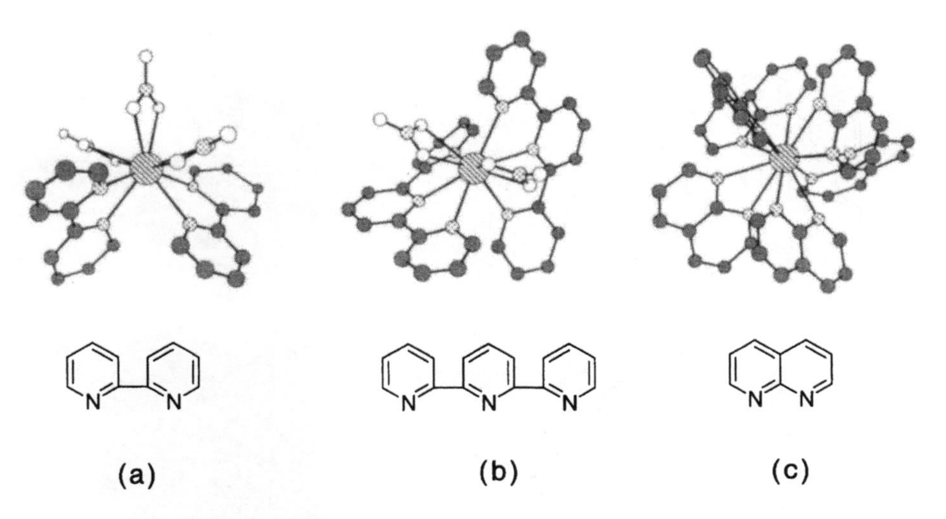

FIGURE 3.21 STRUCTURES OF LA COMPLEXES WITH AROMATIC N-DONOR LIGANDS.

3.4 COMPLEXES WITH NEUTRAL N-DONOR LIGANDS

The coordination chemistry of lanthanides and actinides with simple alkylamines is very limited and will not be considered here. This section will deal only with aromatic N-donors and Schiff base ligands.

3.4.1 Aromatic N-donors

Innumerable complexes of lanthanide and actinide salts with aromatic N-donors such as pyridine, 2,2'-bipyridyl and terpyridyl have been prepared, and many of them have been characterized by X-ray diffraction. These complexes were important in establishing the principles of f-block coordination chemistry, and many previously unknown coordination numbers and geometries were discovered in the course of these investigations. Some examples of complexes with these simple ligands, all of which are prepared in non-aqueous solution, are shown in Figure 3.21. N-methylimidazole is a very strong donor and has been used widely in lanthanide chemistry to stabilize monomeric complexes.

More recent interest in lanthanide complexes containing aromatic N-donors has focused on their photophysical properties and in particular on their use as luminescent probes. This area has been reviewed by Sabbatini, Guardigli and Lehn (1993). The hexadentate sexipyridine ligand, spy, forms dinuclear double helical complexes with transition metal ions but Constable, Chotalia, and Tocher (1992) have found that the large Ln^{3+} ions are able to form mononuclear complexes. The structure of $[Eu(NO_3)_2(spy)]^+$ is shown in Figure 3.22. Although the Eu^{3+} ion can bind to all six N-donors, the spy ligand must adopt a helical conformation. Self-assembly of triple helical Ln_2L_3 complexes has been investigated by Piguet *et al*.(1993). The crystal structure of such a complex is shown in Figure 3.22. L is designed to act as a bis-terdentate ligand, and the requirement of Ln^{3+} for high coordination numbers, combined with weak stacking interactions between ligands results in formation

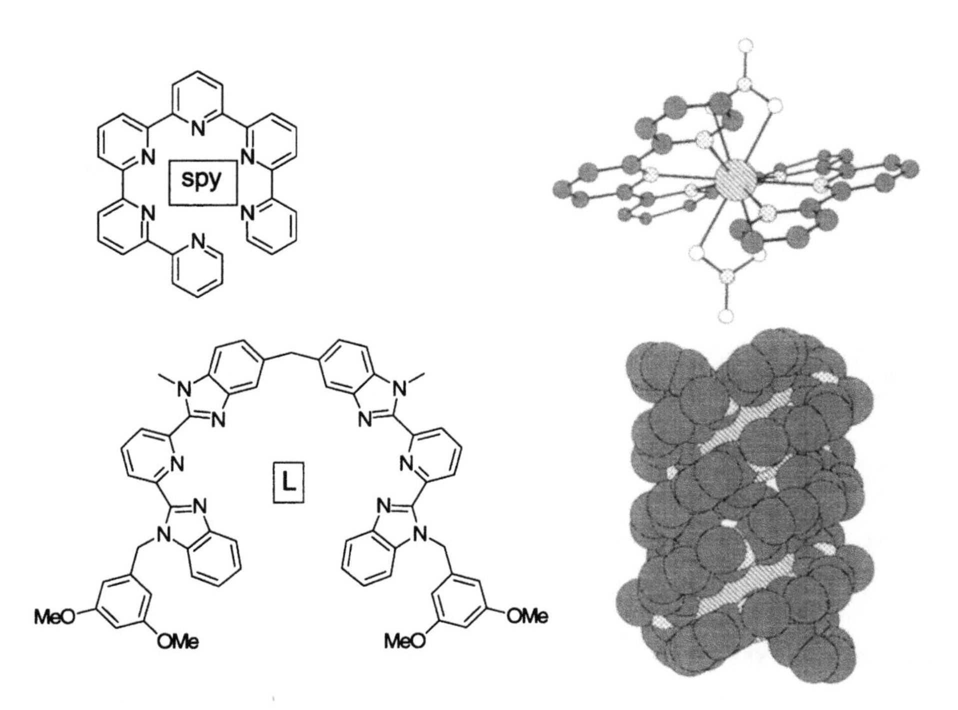

FIGURE 3.22 STRUCTURES OF HELICAL COMPLEXES OF EU.

of a triple helical dinuclear complex with each Ln^{3+} achieving a coordination number of nine.

3.4.2 Schiff base macrocycles

Schiff base macrocycle complexes can usually be prepared easily by template condensation reactions between diamines and dicarbonyl compounds in the presence of the appropriate metal salt, and have been well-studied for both lanthanides and actinides. The size of the macrocycle can be varied by appropriate choice of the organic building blocks, and mononuclear, dinuclear and even trinuclear complexes can be prepared. Schiff base macrocycle chemistry has been reviewed by Guerrio, Tamburini and Vigato (1995) and by Alexander (1995).

The condensation between ethylenediamine and 2,6-diacetylpyridine in the presence of a lanthanide salt was first reported by Backer-Dirks *et al.* (1979) and gives a hexaazamacrocycle complex as shown in Figure 3.23. The condensation can also be carried out using 2,6-diformylpyridine to give an analogous complex. These complexes are of great significance as they are very stable in aqueous solution and even addition of OH^- or F^- does not result in decomplexation of Ln^{3+}. The macrocyclic ligand is not planar, and the distortion from planarity increases with decreasing Ln^{3+} radius. Two views of the $La(NO_3)_3$ complex are shown in Figure 3.23, where the distortion from planarity of the macrocycle can be seen quite clearly. Analogous kinetically inert complexes of UO_2^{2+} can also be prepared by this Schiff base condensation reaction.

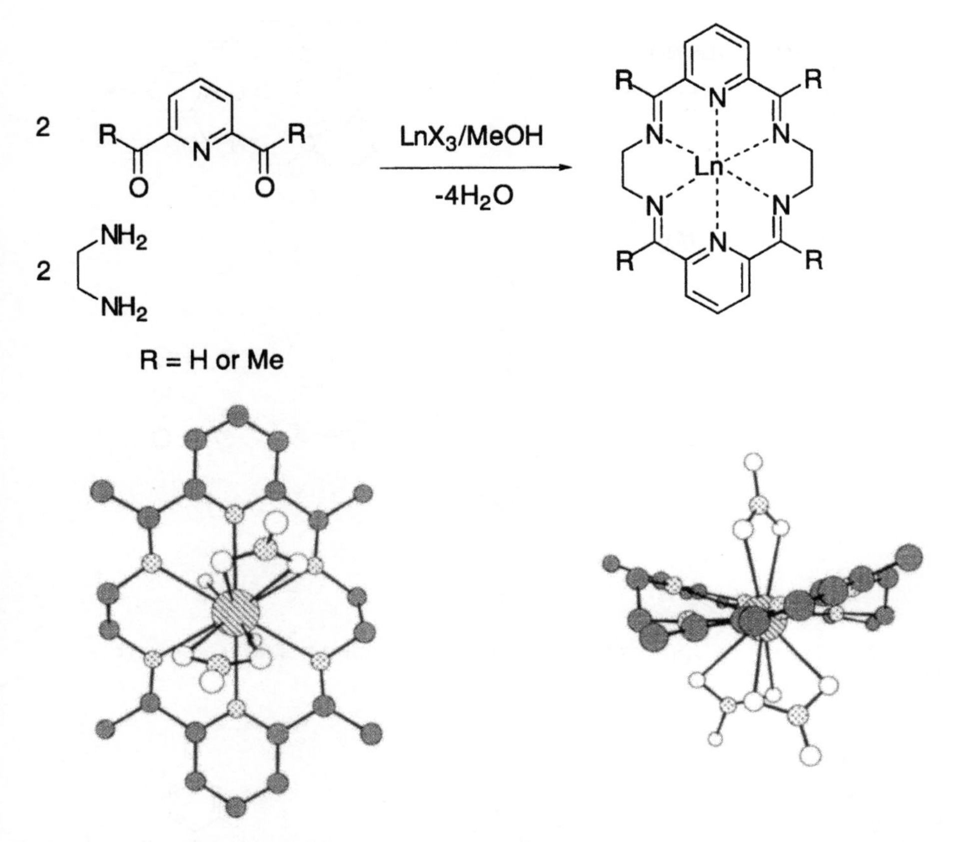

FIGURE 3.23 TEMPLATE SYNTHESIS OF A LANTHANIDE SCHIFF BASE MACROCYCLE COMPLEX, AND TWO VIEWS OF $[LA(NO_3)_3(L)]$.

The size of the macrocyclic cavity can be increased by using a longer chain diamine, and Kahwa *et al.* (1989) reported the preparation of a Gd_2 complex by the Schiff-base condensation of 2,6-diformyl *p*-cresol with a hybrid polyether as shown in Figure 3.24. This was the first example of a homodinuclear lanthanide macrocycle complex and was of particular interest for the study of potential interactions between Ln^{3+} ions.

The Schiff-base condensation reaction of *p*-chloro 2,6-diformyl phenol with diethylenetriamine in the presence of $UO_2(NO_3)_2$ gives a large macrocycle coordinated to one UO_2^{2+} ion as shown in Scheme 3.4. Heterobimetallic complexes can be prepared by addition of salts of transition metals such as Co^{2+}, Ni^{2+} or Cu^{2+}.

- Coordination chemistry with simple alkylamines is very limited.
- Aromatic N-donors form many complexes with Ln and An.
- Complexes of Eu^{3+} and Tb^{3+} with pyridine-containing donors are often strongly luminescent.
- Schiff base macrocycle complexes of Ln and An are readily prepared by template condensation reactions.

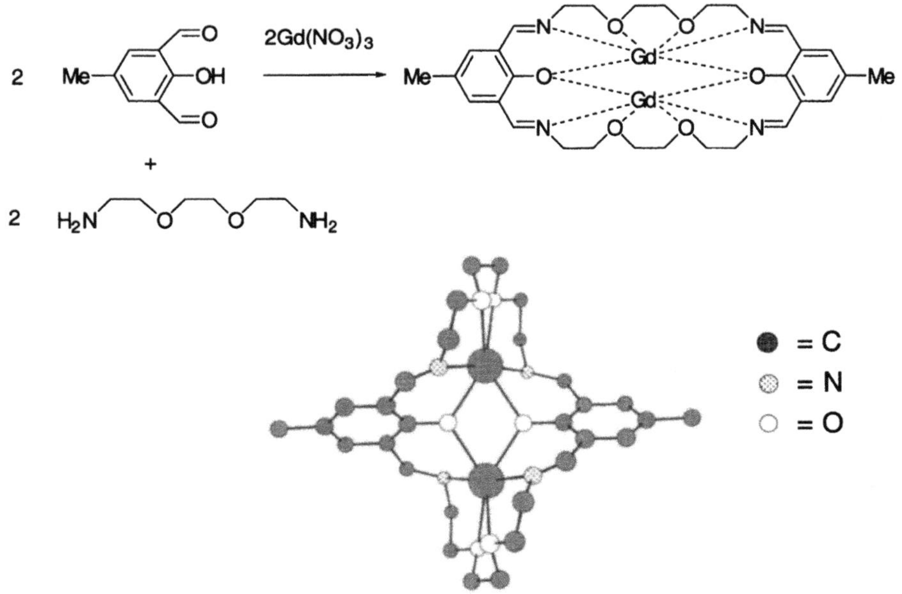

FIGURE 3.24 SYNTHESIS AND STRUCTURE OF A Gd_2 SCHIFF BASE MACROCYCLE COMPLEX.

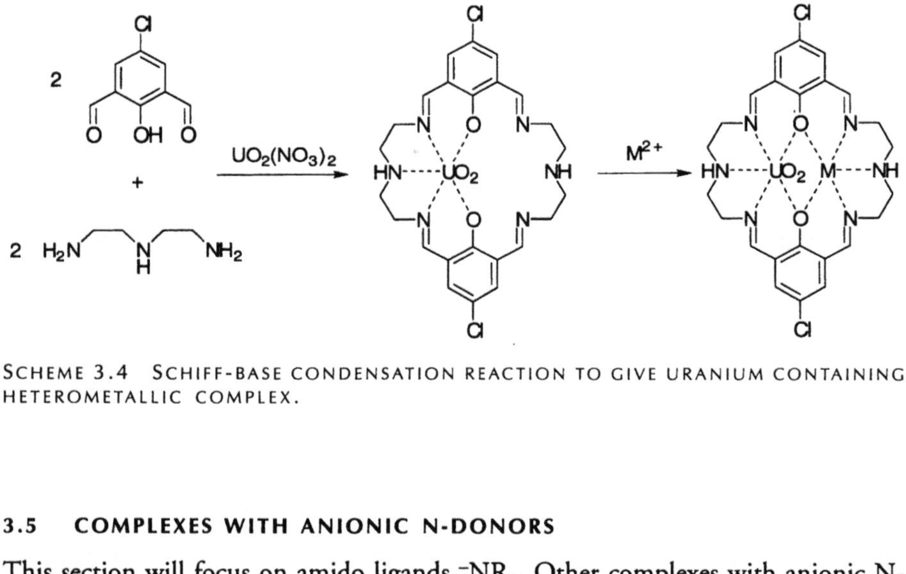

SCHEME 3.4 SCHIFF-BASE CONDENSATION REACTION TO GIVE URANIUM CONTAINING HETEROMETALLIC COMPLEX.

3.5 COMPLEXES WITH ANIONIC N-DONORS

This section will focus on amido ligands $^{-}NR_2$. Other complexes with anionic N-donors are also well known *e.g.* N-bonded SCN^-, and complexes with porphyrins and phthalocyanins.

3.5.1 Hexamethyldisilylamide Complexes

Lanthanides

The hexamethyldisilylamide ($N(SiMe_3)_2$) ligand, often simply referred to as 'silylamide' has played an important role in the history of f-element chemistry. It is a very bulky

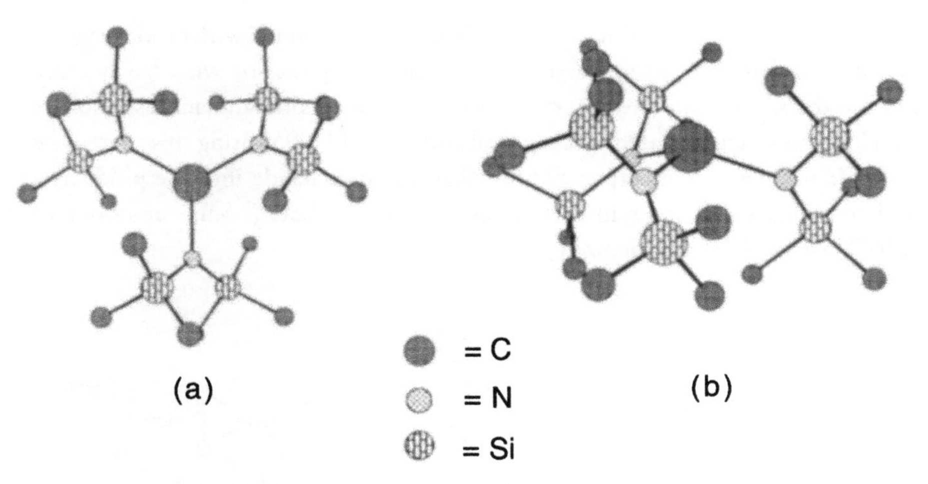

FIGURE 3.25 STRUCTURE OF $[Eu\{N(SiMe_3)_2\}_3]$ (A) VIEW ALONG C_3 AXIS (B) VIEW PERPENDICULAR TO C_3 AXIS.

ligand and has been used to prepare low-coordinate complexes of transition metals. In 1972 the first lanthanide tris-silylamides $[Ln\{N(SiMe_3)_2\}_3]$ were reported, at a time when a coordination number of 6 was considered low for the large Ln^{3+} ions. Molecular weight determination by measurement of boiling-point elevation in benzene showed that the complexes were monomeric and therefore 3-coordinate. Definitive proof of the structure came with the X-ray crystal structure determination of $[Eu\{N(SiMe_3)_2\}_3]$ by Ghotra, Hursthouse and Welch (1973). All $[Ln\{N(SiMe_3)_2\}_3]$ for which structures have been determined show a pyramidal geometry with Ln slightly out of the plane of the three N atoms; the silylamide ligands adopt a propellor arrangement giving the complex overall C_3 symmetry. The crystal structure of $[Eu\{N(SiMe_3)_2\}_3]$ is shown in Figure 3.25. In solution the complexes have zero dipole-moment and are thus planar; the pyramidalization in the solid state has been attributed to attractive intramolecular van der Waals interactions between the ligands.

Lanthanide tris-silylamides are prepared in high yield by the reaction of anhydrous $LnCl_3$ with $LiN(SiMe_3)_2$ in THF under rigorously anhydrous conditions. They are very soluble in organic solvents including pentane, they are reasonably volatile, subliming at about 100°C at 10^{-4} mmHg, and they can be crystallised easily to give highly pure materials. As expected, the bonding in $[Ln\{N(SiMe_3)_2\}_3]$ is essentially ionic, and their reactivity is dominated by protonation reactions which eliminate $HN(SiMe_3)_2$; rapid decomposition occurs in the presence of even traces of water.

Although the silylamide ligand is very bulky, $[Ln\{N(SiMe_3)_2\}_3]$ are still slightly coordinatively unsaturated and reversible adduct formation occurs with several Lewis bases such as THF, Bu^tCN, Bu^tNC. The chemistry with Ph_3PO is rather more complicated and intriguing: addition of one equivalent of Ph_3PO results in simple adduct formation, but addition of two equivalents of Ph_3PO leads to formation of a unique dimeric complex $[Ln_2\{N(SiMe_3)_2\}_4(O_2)(Ph_3PO)_2]$ which contains a μ_2-peroxo ligand, presumably derived from Ph_3PO. The relative ease of synthesis and purification of $[Ln\{N(SiMe_3)_2\}_3]$, and their ready reaction of with protic reagents have been exploited in the synthesis of many new complexes by what has become known as the 'silylamide route.' For example the first homoleptic lanthanide selenolates and tellurolates were prepared by Cary, Ball and Arnold (1995) using this route. The silylamide route avoids incorporation of alkali metal or halide into the product, a problem which often arises in metathesis reactions of $LnCl_3$. Some reactions of $[Ln\{N(SiMe_3)_2\}_3]$ are summarized in Scheme 3.5.

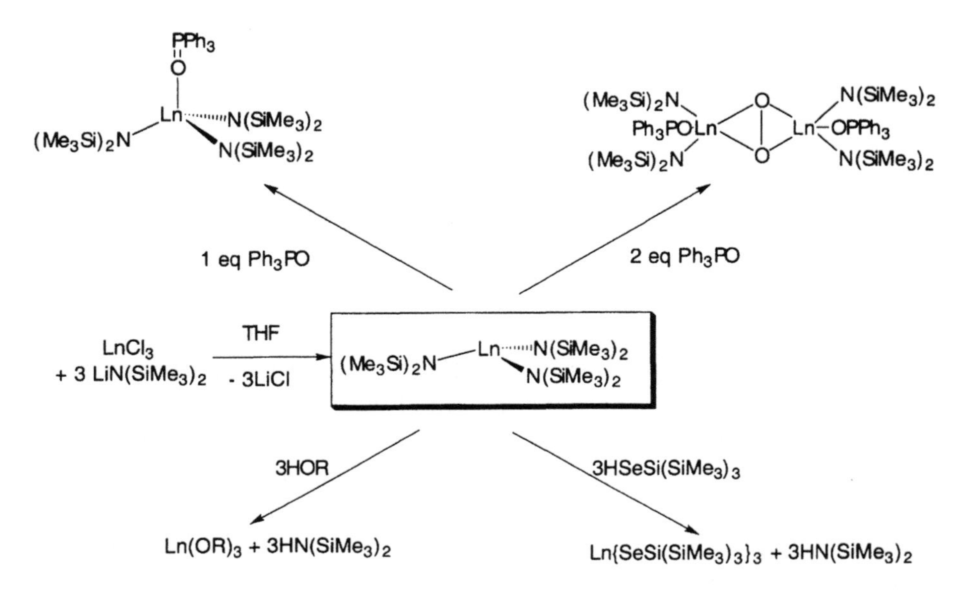

SCHEME 3.5 REACTIONS OF LANTHANIDE TRIS(SILYLAMIDES).

Complexes of Sm(II), Eu(II) and Yb(II) with the silylamide ligand are also known. They are best prepared by metathesis reactions of LnI_2 with $NaN(SiMe_3)_2$. Because of the much larger size of Ln(II) ions the $Ln\{N(SiMe_3)_2\}_2$ unit is very coordinatively unsaturated and exists only in the presence of donor ligands in complexes such as $[Sm\{N(SiMe_3)_2\}_2(THF)_2]$ reported by Evans, Drummond and Zhang (1988) or $[Yb\{N(SiMe_3)_2\}_2(Me_2PCH_2CH_2PMe_2)]$, the first structurally characterised example of a lanthanide complex with a tertiary phosphine ligand, which was reported by Tilley, Andersen and Zalkin (1982). The crystal structures of both these complexes are shown in Figure 3.26.

Actinides

Actinide silylamide chemistry began with complexes in the readily available +4 oxidation state; $[ThCl\{N(SiMe_3)_2\}_3]$ was prepared by Bradley, Ghotra and Hart (1974) by the reaction of $ThCl_4$ with 4 equivalents of $LiN(SiMe_3)_2$. Steric reasons prevent the formation of homoleptic An(IV) silylamides: attempts to prepare $[An\{N(SiMe_3)_2\}_4]$ by the reaction of $[AnCl\{N(SiMe_3)_2\}_3]$ with $NaN(SiMe_3)_2$

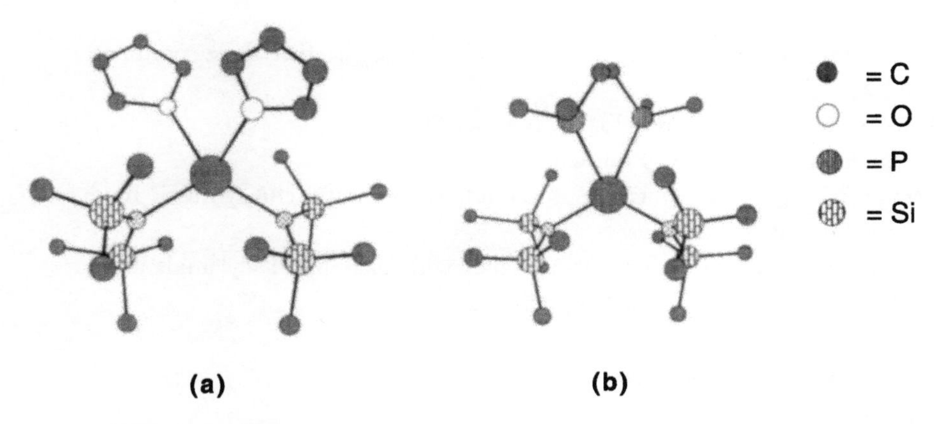

FIGURE 3.26 STRUCTURES OF LANTHANIDE(II) SILYLAMIDES (A) [SM{N(SIME3)2(THF)2] (B) [YB{N(SIME3)2(DMPE)].

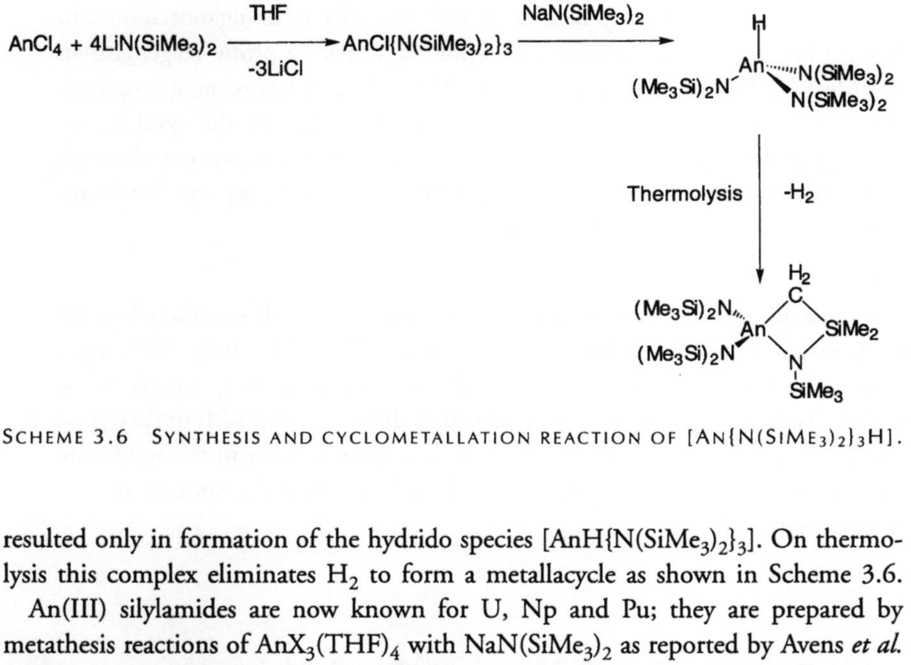

$$AnCl_4 + 4LiN(SiMe_3)_2 \xrightarrow[-3LiCl]{THF} AnCl\{N(SiMe_3)_2\}_3 \xrightarrow{NaN(SiMe_3)_2} [AnH\{N(SiMe_3)_2\}_3]$$

H
(Me3Si)2N–An–N(SiMe3)2, N(SiMe3)2

Thermolysis | $-H_2$

H_2C, SiMe2, N, SiMe3
(Me3Si)2N, (Me3Si)2N–An

SCHEME 3.6 SYNTHESIS AND CYCLOMETALLATION REACTION OF [AN{N(SIME3)2}3H].

resulted only in formation of the hydrido species $[AnH\{N(SiMe_3)_2\}_3]$. On thermolysis this complex eliminates H_2 to form a metallacycle as shown in Scheme 3.6.

An(III) silylamides are now known for U, Np and Pu; they are prepared by metathesis reactions of $AnX_3(THF)_4$ with $NaN(SiMe_3)_2$ as reported by Avens *et al.* (1994) and are isostructural with their lanthanide analogues. Like their lanthanide analogues they are susceptible to protonolysis. However, unlike their lanthanide analogues they do not form adducts with Lewis bases, and they are susceptible to oxidation: an attempt by Andersen (1979) to form an adduct between $[U\{N(SiMe_3)_2\}_3]$ and Me_3NO resulted in oxidation to $[UO\{N(SiMe_3)_2\}_3]$ as shown in Scheme 3.7.

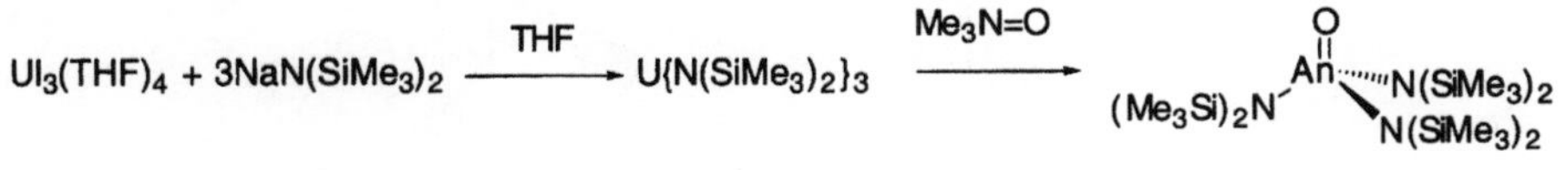

$$UI_3(THF)_4 + 3NaN(SiMe_3)_2 \xrightarrow{THF} U\{N(SiMe_3)_2\}_3 \xrightarrow{Me_3N=O} [An(=O)\{N(SiMe_3)_2\}_3]$$

SCHEME 3.7 SYNTHESIS OF [U{N(SIME3)2}3] AND ITS REACTION WITH ME3N=O.

- $[Ln\{N(SiMe_3)_2\}_3]$ are known for all Ln. They were the first three-coordinate complexes of Ln.
- $[An\{N(SiMe_3)_2\}_3]$ are known for U, Np and Pu.
- Ln and An silylamides are Brønsted bases.
- Reaction with protic reagents (the 'silylamide route') is a useful synthetic route to other Ln and An complexes.
- Synthesis is by metathesis of Ln or An halide with alkali metal silylamide under strictly anhydrous and anaerobic conditions.

3.5.2 Dialkylamide Complexes

Lanthanides

The chemistry of lanthanides with dialkylamide ligands NR_2 is much more limited than that with silylamides. Lanthanide tris-di-isopropylamides are known, but only as their THF adducts $[Ln(NPr^i_2)_3(THF)]$. The smaller size of the di-isopropylamide ligand and its increased basicity compared with the silylamide ligands results in $[Ln(NPr^i_2)_3(THF)]$ being even more susceptible to attack by protic reagents than $[Ln\{N(SiMe_3)_2\}_3]$. Reactions of $[Ln(NPr^i_2)_3(THF)]$ with secondary organophosphines have been exploited by Aspinall, Moore and Smith (1992) in the synthesis of lanthanide phosphido complexes. The smaller size of NPr^i_2 also allows coordination of 4 ligands in the 'ate' complexes $[Li(THF)][Ln(\mu\text{-}Pr^i_2)_2(NPr^i_2)_2]$ reported by Evans, Anwander and Ziller (1995) and shown in Figure 3.27.

Actinides

The less sterically less demanding dialkylamide ligands are well established in Th and U chemistry. $[An(NEt_2)_4]$ have been known for Th and U since 1979. They are prepared by reaction of $AnCl_4$ with $LiNEt_2$ and are sufficiently volatile to be purified by distillation. An X-ray crystal structure determination of $[U(NEt_2)_4]$ by Reynolds, Zalkin and Templeton (1976) showed it to be a dimer in the solid state with two $\mu_2\text{-}NEt_2$ groups and approximate trigonal bipyramidal geometry at U as shown in Scheme 3.8.

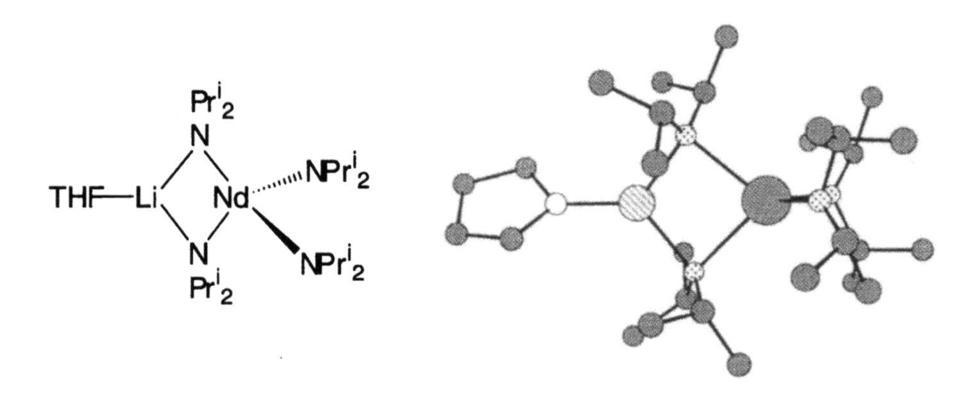

FIGURE 3.27 STRUCTURE OF $[Li(THF)][Ln(\mu\text{-}NPr^i_2)_2(NPr^i_2)_2]$.

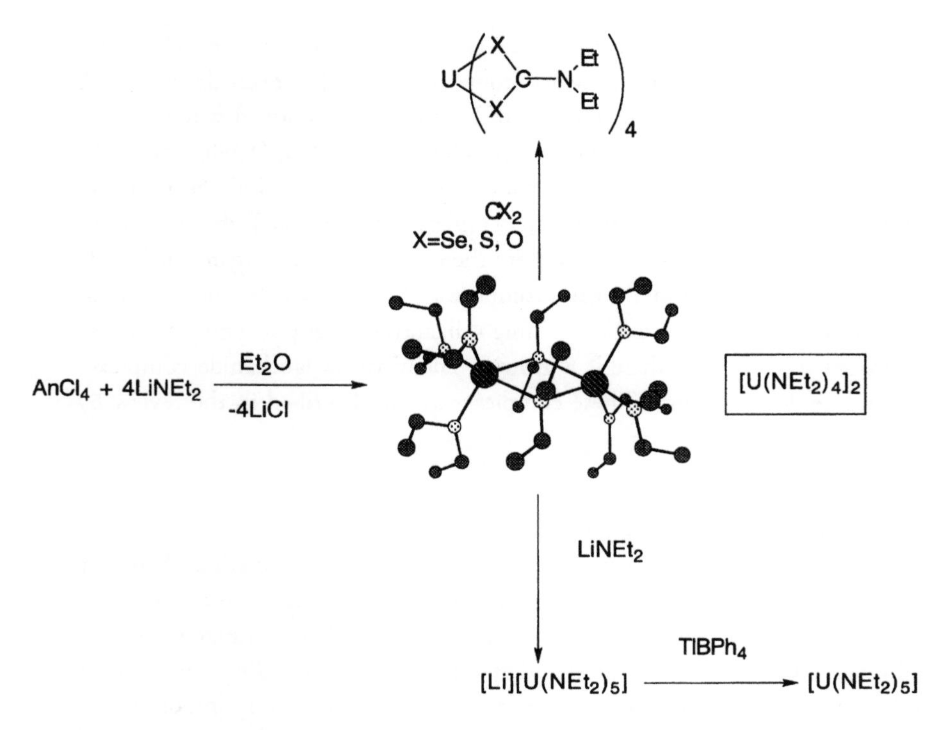

SCHEME 3.8 SYNTHESIS AND REACTIONS OF $[U(NEt_2)_4]_2$.

Addition of $LiNEt_2$ to $[U(NEt_2)_4]$ leads to the five-coordinate 'ate' complex $[Li][U(NEt_2)_5]$ which can be oxidized with $TlBPh_4$ to produce the U(V) complex $[U(NEt_2)_5]$ as reported by Berthet and Ephritikhine (1993). An increase in ligand size leads to monomeric structures for $[An(NPh_2)_4]$. Reaction chemistry of actinide dialkylamides is again dominated by protonolysis of the An-N bond, which has been exploited in the synthesis of new actinide complexes. Insertion of CX_2 (X = O, S, Se) into the An-N bond to form carbamates has also been demonstrated. Some reactions of $[U(NEt_2)_4]_2$ are summarized in Scheme 3.8.

- NR_2^- complexes are stronger Brønsted bases than silylamide complexes.
- NR_2^- are less sterically demanding than $N(SiMe_3)_2^-$. Higher coordination numbers are observed.
- Protonolysis of An-N bonds and insertion into An-N bonds are well known.

3.6 COMPLEXES WITH HEAVIER DONOR ATOMS

Lanthanide and actinide ions generally behave as typical hard Lewis acids and so, although the heavier halides are very well known, for many years it was assumed that these elements would not form complexes with 2nd row (or heavier) donors such as P and S. It is true that, given the choice, the f-block elements will preferentially coordinate to hard donors, but they will form complexes with soft donors if there is no alternative, and since the mid-1980's, a rich coordination chemistry of f-block elements with P and heavier Group 16 donors has been developed. A review by Nief (1998) describes complexes of Ln and An with non-first row donors (excluding halogens) and a review by Fryzuk, Haddad and Berg (1990) covers the coordination chemistry of Ln and An with neutral P donors. This Section will concentrate mainly on simple 'Inorganic' complexes with S and P donor ligands. In addition to the complexes described here there are numerous organolanthanide and -actinide phosphido and thiolato complexes, most of which contain Cp or related ligands. Complexes with P-containing π-ligands such as phospholyl have also been prepared recently, and there is a growing chemistry of lanthanide complexes with Se and Te donor ligands. These complexes are all described in the review by Nief.

3.6.1 Complexes with P-donor Ligands

Complexes with Neutral P-donors

Neutral P-donor ligands have played a central role in the coordination chemistry of the d-transition metals, especially those in the middle or late in the series, and π-back-bonding is an important feature, especially in low valent complexes. Because of the limited radial extension of the 4f and 5f orbitals, π-back-bonding cannot occur for lanthanide complexes with phosphines, and is unlikely to be very important for the actinides. Complexes of phosphines with f-block elements are therefore expected to be much less stable than those of d-transition metals.

Lanthanides

The first evidence for phosphine coordination to a lanthanide was reported in 1965. Addition of PPh_3 to Cp_3Yb resulted in a change in the electronic absorption spectrum which was interpreted as evidence for adduct formation. The adduct $[Cp_3Yb(PPh_3)]$ was subsequently isolated and characterized by elemental analysis. A small number of related complexes were isolated using different phosphine ligands, but the chemistry was often complicated and mixtures were frequently obtained. The first fully characterized lanthanide phosphine complex was prepared by addition of the chelating diphosphine dmpe to $[Yb\{N(SiMe_3)_2\}_2(OEt_2)_2]$ to give $[Yb\{N(SiMe_3)_2\}_2(dmpe)_2]$. Hybrid chelating P,O donors (Hitchcock, Lappert and MacKinnon (1988)) and P,N donors (Fryzuk, Haddad and Rettig (1992)) have also been used to prepare complexes of the lanthanides with tertiary phosphine donor groups. Examples of the preparation of lanthanide complexes with tertiary phosphine donors are shown in Scheme 3.9. NMR spectroscopy of Y complexes shows a ^{31}P-^{89}Y coupling of 52 Hz for the P,N complex and 59 Hz for the P,O complex. As expected the Ln-phosphine bonds are rather labile.

Actinides

Cp_3An, like their lanthanide counterparts, form adducts with monodentate tertiary phosphines. In a comparative study of $[(MeCp)_3M\{P(OCH_2)_3CEt\}]$ and $[(MeCp)_3M(quinuclidine)]$ (M=Ce or U), Brennan *et al.* (1988) found that the U-P bond was significantly shorter (2.988 Å) than the Ce-P bond (3.086 Å), whereas

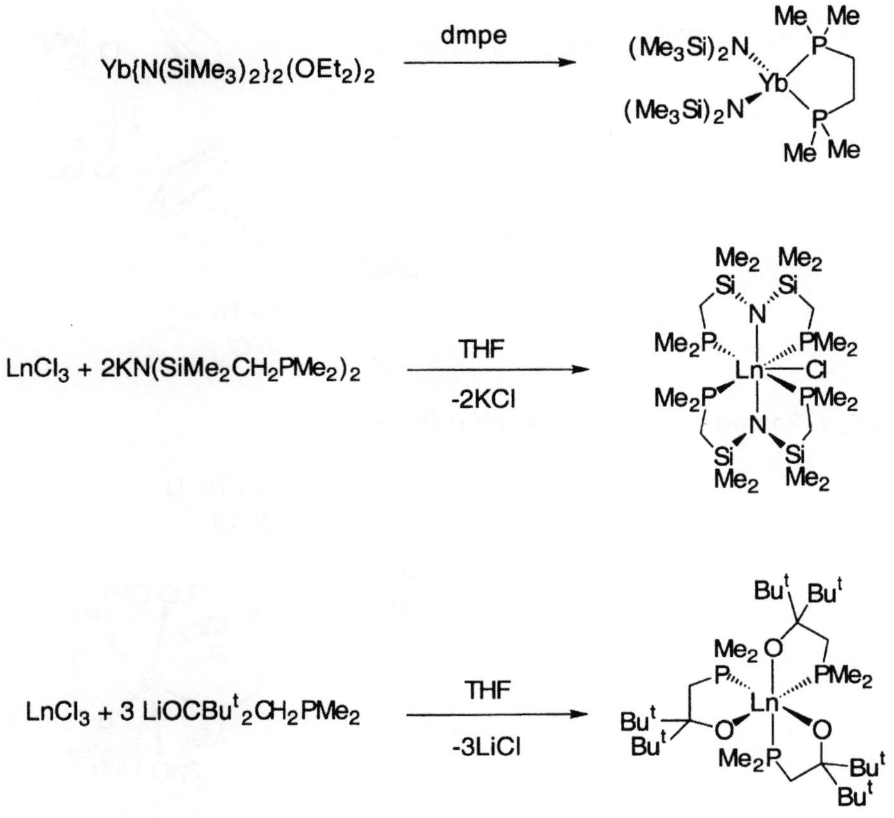

$Yb\{N(SiMe_3)_2\}_2(OEt_2)_2$ $\xrightarrow{\text{dmpe}}$ $Yb\{N(SiMe_3)_2\}_2(dmpe)$

$LnCl_3 + 2KN(SiMe_2CH_2PMe_2)_2$ $\xrightarrow[\text{-2KCl}]{\text{THF}}$

$LnCl_3 + 3\ LiOCBu^t{}_2CH_2PMe_2$ $\xrightarrow[\text{-3LiCl}]{\text{THF}}$

SCHEME 3.9 SYNTHESIS OF LANTHANIDE COMPLEXES WITH PHOSPHINE LIGANDS.

the U-N (2.764 Å) and the Ce-N (2.789 Å) distances were very similar, as expected for purely electrostatic bonding (Ce(III) and U(III) have very similar ionic radii). The decrease in M-P distances from Ce to U has been interpreted as evidence for some π-back-bonding in the U-P interaction.

Complexes of UCl_4 with tertiary phosphines have been reported since the 1960's, however many of these compounds were not well characterized and some have subsequently been shown to contain phosphine oxides or phosphonium salts. The first well characterized phosphine complexes were reported by Edwards, Andersen and Zalkin (1981). They have the general formula $[MX_4(dmpe)_2]$ where M=Th, U and X=Cl, Br. These complexes are prepared in CH_2Cl_2 or in neat dmpe at elevated temperatures. They are hydrocarbon soluble, quite stable, and can be used as starting materials for An chemistry as shown in Scheme 3.10. The U complexes sublime intact under high vacuum. Although $[AnX_4(dmpe)_2]$ have not been characterized crystallographically, they are believed on the basis of their physical properties to be eight-coordinate monomers. Hybrid amido phosphine donors have also been used in actinide chemistry by Coles *et al.* (1995).

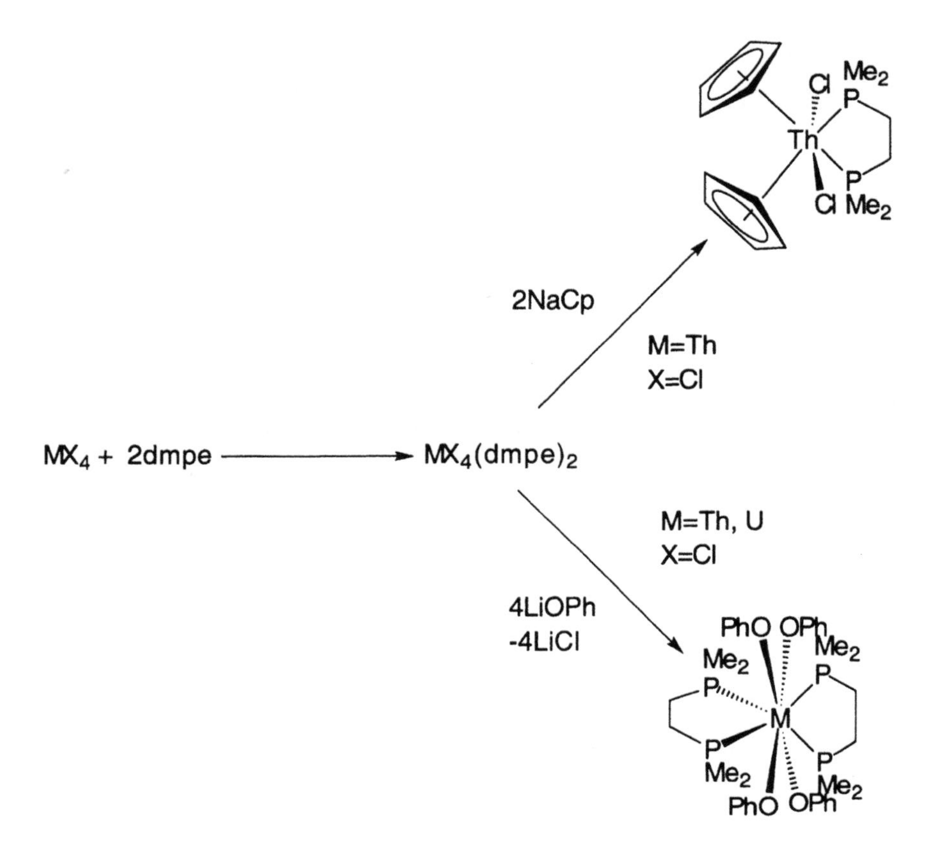

SCHEME 3.10 SYNTHESIS AND REACTIONS OF ACTINIDE COMPLEXES WITH PHOSPHINE LIGANDS.

Complexes with Phosphido ligands

Lanthanides

Lanthanide complexes with phosphido ligands were first prepared by Schumann *et al.* (1986) and by Aspinall, Moore and Smith (1992) *via* protonolysis reactions of reactive lanthanide precursors with the relatively acidic Ph_2PH as shown in Scheme 3.11. Yb(II) Sm(II) phosphido complexes have also been obtained by Rabe, Yap and Rheingold (1995) *via* protonolysis reactions of Ln(II) silylamides with Ph_2PH.

Metathesis reactions of lanthanide halides or triflates with alkali metal phosphides have been used to obtain complexes with the bulky phosphides $PBu^t{}_2{}^-$ and $P(SiMe_3)_2{}^-$ as shown in Scheme 3.12. Reactions with $LiPBu^t{}_2$ are dependent on the identity of Ln: Rabe, Riede and Schier (1996a) have shown that the large $LaCl_3$ gives a 4-coordinate 'ate' complex whereas smaller Ln give neutral $[Ln(PBu^t{}_2)_3(THF)_n]$. $Ln(OTf)_3$ where Ln=Sm, Eu or Yb are reduced by $LiPBu^t{}_2$ to give 4-coordinate Ln(II) 'ate' complexes. Metathesis reactions with $KP(SiMe_3)_2$ have been used by Rabe, Riede and Schier (1996b) to give neutral Ln(II) and Ln(III) phosphido complexes. In structurally characterized complexes with terminal phosphido ligands the geometry around P is found to be either planar or very close to planar.

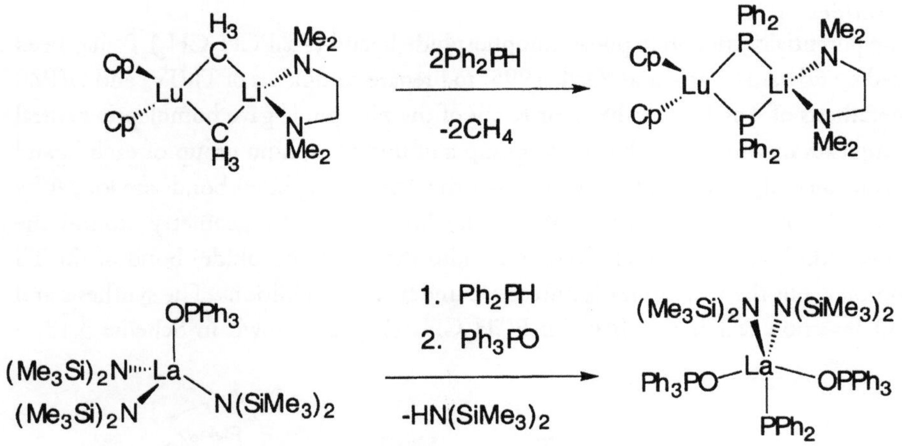

$Cp_2Lu(\mu-CH_3)_2Li(TMEDA)$ + 2Ph2PH → $Cp_2Lu(\mu-PPh_2)_2Li(TMEDA)$; −2CH4

H3 C Me2 N Cp Lu Li Cp C H3 N Me2 — 2Ph2PH, -2CH4 → Ph2 P Me2 N Cp Lu Li Cp P Ph2 N Me2

OPPh3 (Me3Si)2N La (Me3Si)2N N(SiMe3)2 — 1. Ph2PH 2. Ph3PO, -HN(SiMe3)2 → (Me3Si)2N N(SiMe3)2 La Ph3PO OPPh3 PPh2

SCHEME 3.11 SYNTHESIS OF LANTHANIDE COMPLEXES WITH PHOSPHIDO LIGANDS BY PROTONOLYSIS REACTIONS WITH 2° PHOSPHINES.

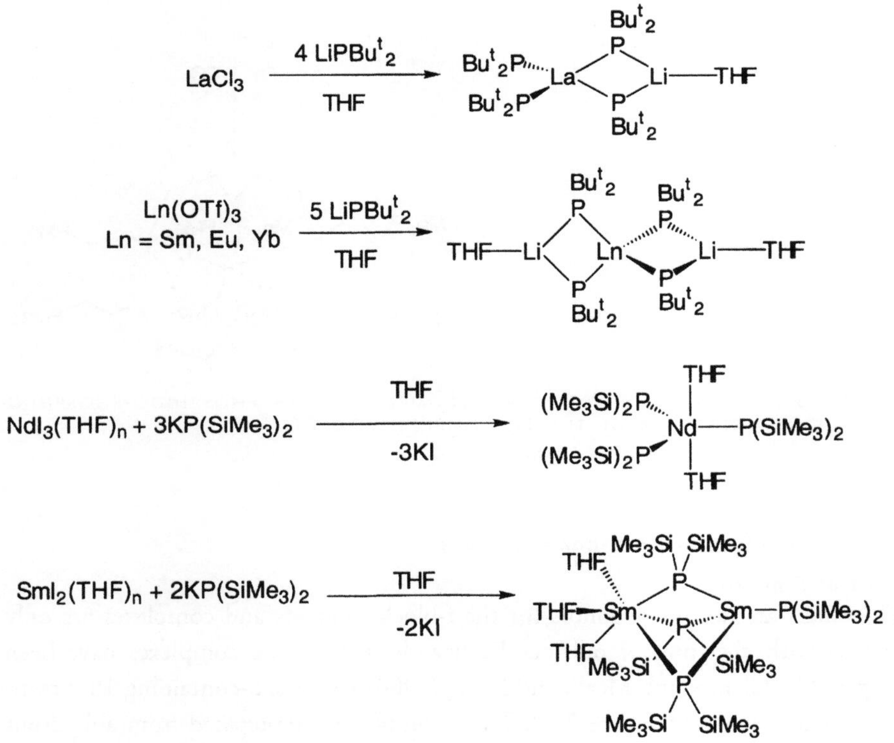

LaCl3 — 4 LiPBu$^t_2$, THF → Bu$^t_2$ P Bu$^t_2$P La Li—THF Bu$^t_2$P P Bu$^t_2$

Ln(OTf)3 Ln = Sm, Eu, Yb — 5 LiPBu$^t_2$, THF → Bu$^t_2$ P Bu$^t_2$ P THF—Li Ln Li—THF P P Bu$^t_2$ Bu$^t_2$

NdI3(THF)n + 3KP(SiMe3)2 — THF, -3KI → THF (Me3Si)2P Nd—P(SiMe3)2 (Me3Si)2P THF

SmI2(THF)n + 2KP(SiMe3)2 — THF, -2KI → Me3Si SiMe3 THF P THF—Sm P Sm—P(SiMe3)2 THF Me3Si SiMe3 P Me3Si SiMe3

SCHEME 3.12 SYNTHESIS OF LANTHANIDE COMPLEXES WITH PHOSPHIDO LIGANDS BY METATHESIS REACTIONS.

Actinides

The potentially tridentate phosphinophosphide ligand $(Ph_2PCH_2CH_2)_2P^-$ has been used by Edwards, Parry and Read (1995) to prepare complexes of Th(IV) and U(IV). Metathesis of $AnCl_4$ with the Li or K salt of the phosphide gave homoleptic neutral complexes in which the phosphido group and one phosphino group of each ligand is coordinated. X-ray studies showed that the An-P(phosphine) bonds are longer by about 20 pm than the An-P (phosphide) bonds, and the geometry around the phosphido P atom is planar. CO inserts into the An-P(phosphide) bond of the Th complex, but the U complex is unreactive under these conditions. The synthesis and CO insertion reaction of $[An\{(Ph_2PCH_2CH_2)_2P\}_4]$ are shown in Scheme 3.13.

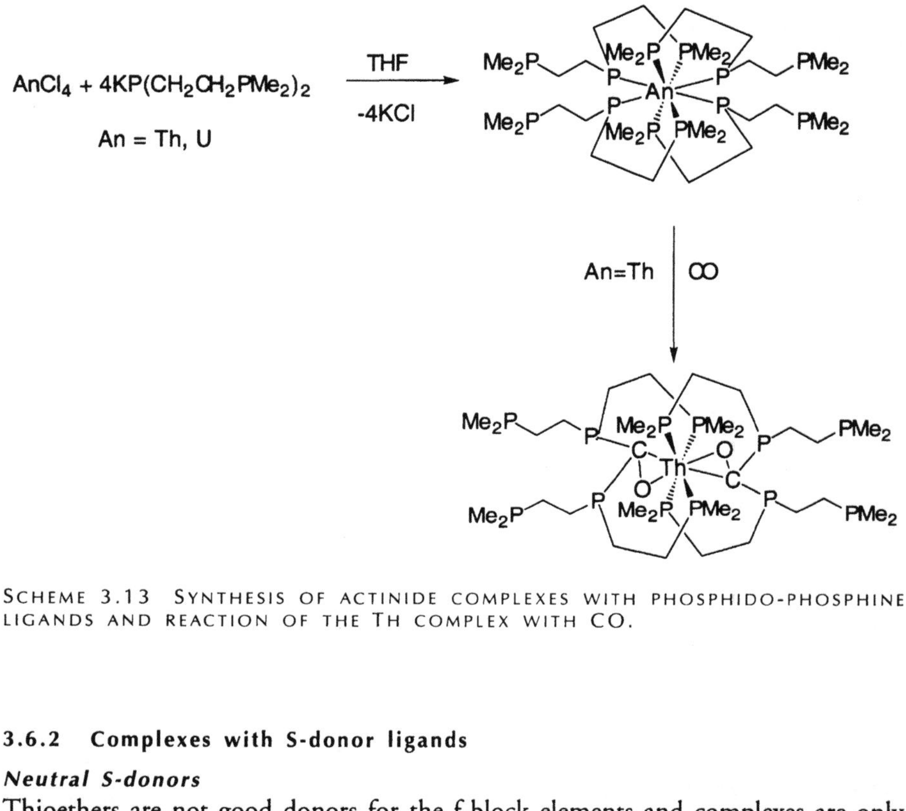

SCHEME 3.13 SYNTHESIS OF ACTINIDE COMPLEXES WITH PHOSPHIDO-PHOSPHINE LIGANDS AND REACTION OF THE TH COMPLEX WITH CO.

3.6.2 Complexes with S-donor ligands

Neutral S-donors

Thioethers are not good donors for the f-block elements and complexes are only formed with chelating or macrocyclic ligands. Lanthanide complexes have been prepared by Ciampolini, Mealli and Nardi (1980) with the S-containing 18-crown-6 analogues shown in Figure 3.27. These complexes are prepared from anhydrous lanthanide perchlorates in non-coordinating solvents and they are all extremely hygroscopic, decomposing within minutes to form hydrated lanthanide perchlorates and free macrocycle on exposure to moist air. Only the complex of Nd with 18-C-6-S_2 has been characterized crystallographically; the structure of this complex is shown in Figure 3.27.

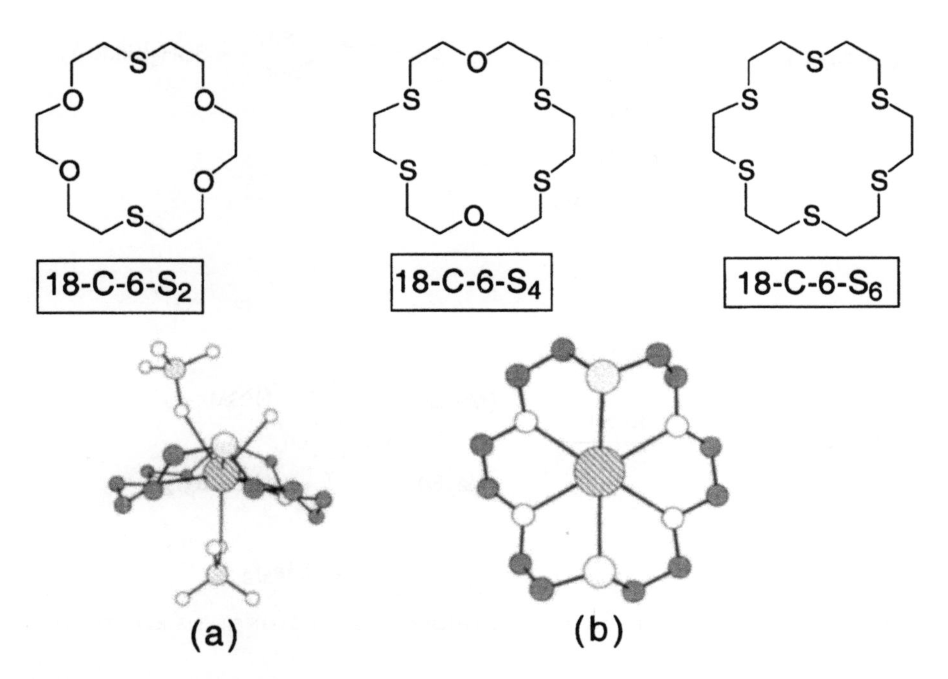

FIGURE 3.28 S-CONTAINING ANALOGUES OF 18-C-6 WHICH FORM COMPLEXES WITH Ln^{3+}. (A) $[La(18\text{-}C\text{-}6\text{-}S_2)(H_2O)(ClO_4^-)_2]^+$ (B) AS (A) WITH H_2O AND ClO_4^- OMITTED FOR CLARITY.

Thiolates

The f-block elements have a rich chemistry with thiolate (SR^-) ligands. Much of this chemistry has been driven by the potential of lanthanide thiolates as precursors for the synthesis of lanthanide sulfides, which have applications as pigments and as optical and electronic materials. The applications of rare earth sulfides have been reviewed by Kumta and Risbud (1994).

There are three main synthetic routes to f-block thiolates. Protonolysis of a reactive complex such as an alkyl or silylamide has the advantage of being applicable to all metals and giving easily separated by-products. Two examples of this type of reaction are shown in Scheme 3.14.

Metathesis reactions are not widely used in the preparation of f-block thiolates, however the first homoleptic six-coordinate lanthanide thiolate was prepared by the metathesis of $YbCl_3$ with $LiSBu^t$ as shown in Scheme 3.15. Reaction of UCl_4 with 6 equiv of NaSR also leads to a six-coordinate 'ate' complex $[\{(THF)_3Na(\mu\text{-}SR)_3\}_2U]$ where R=Bu^t or Ph.

For Sm, Eu and Yb, oxidative addition of RSSR to either the metal or a mercury almalgam of the metal results in clean formation of thiolate complexes which can be isolated as crystals in the presence of neutral donors such as pyridine. The oxidation state of the product depends on the stoichiometry: reaction of Ln with 1 equiv of RSSR gives a Ln(II) product whereas a Ln(III) product is obtained on reaction with 1.5 equiv of RSSR. The donor strength of the neutral ligand also affects the nature of the product:

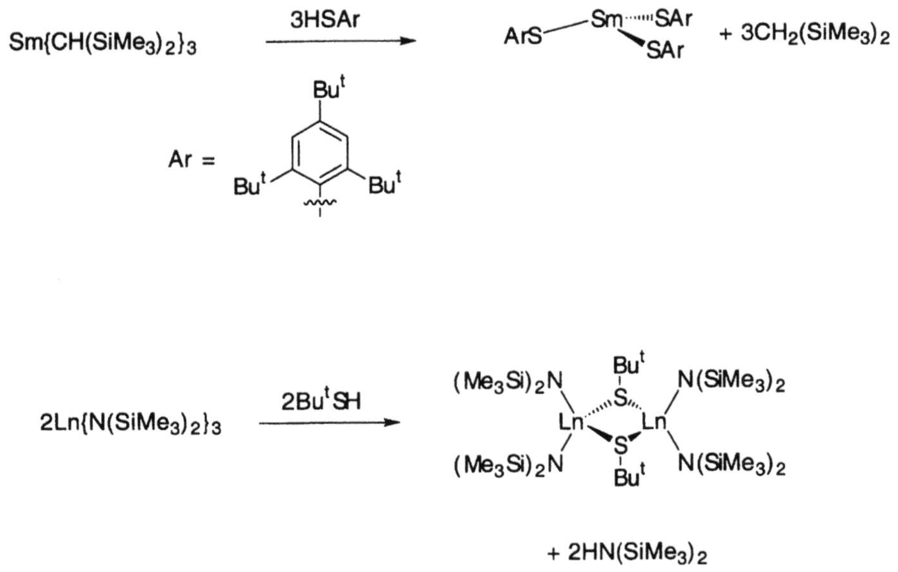

$Sm\{CH(SiMe_3)_2\}_3 \xrightarrow{3HSAr} Sm(SAr)_3 + 3CH_2(SiMe_3)_2$

Ar = 2,4,6-$Bu^t_3C_6H_2$

$2Ln\{N(SiMe_3)_2\}_3 \xrightarrow{2Bu^tSH} [\{(Me_3Si)_2N\}_2Ln(\mu\text{-}SBu^t)]_2 + 2HN(SiMe_3)_2$

SCHEME 3.14 SYNTHESIS OF LANTHANIDE THIOLATES BY PROTONOLYSIS REACTIONS.

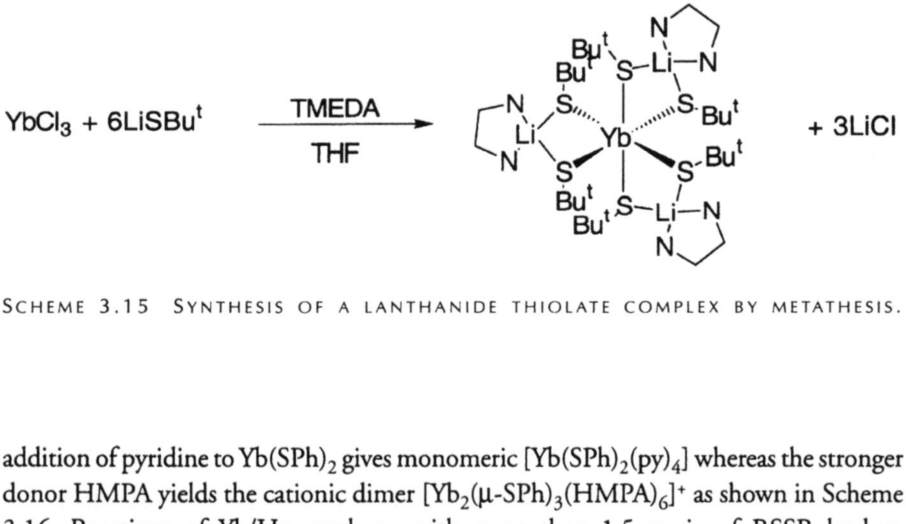

$YbCl_3 + 6LiSBu^t \xrightarrow[THF]{TMEDA}$ [Yb complex with six SBu^t ligands and three Li(TMEDA) units] $+ 3LiCl$

SCHEME 3.15 SYNTHESIS OF A LANTHANIDE THIOLATE COMPLEX BY METATHESIS.

addition of pyridine to $Yb(SPh)_2$ gives monomeric $[Yb(SPh)_2(py)_4]$ whereas the stronger donor HMPA yields the cationic dimer $[Yb_2(\mu\text{-}SPh)_3(HMPA)_6]^+$ as shown in Scheme 3.16. Reactions of Yb/Hg amalgam with more than 1.5 equiv of RSSR lead to formation of cluster complexes which incorporate Hg.

SR^- are very basic ligands and, except when R is very bulky, they have a strong tendency to act as bridging ligands; in the absence of strongly coordinating donors, $Ln(SR)_3$ and $Ln(SR)_2$ are generally insoluble polymeric solids. As expected, all of these complexes are extremely moisture sensitive, reacting with traces of water to produce free thiol and metal oxide or hydroxide. Much interest in lanthanide thiolates has been driven by interest in their thermal decomposition to give lanthanide sulfide species, which otherwise must be prepared using CS_2 or H_2S at high pressures and temperatures.

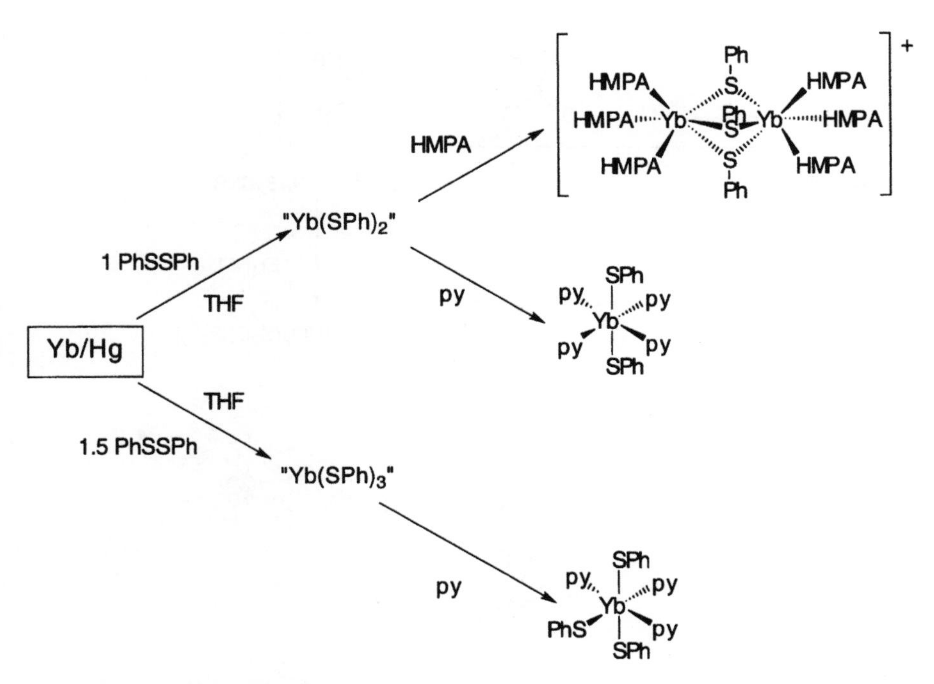

SCHEME 3.16 SYNTHESIS OF LANTHANIDE THIOLATES BY OXIDATIVE ADDITION OF RSSR TO LN.

- The Ln-thiolate bond is not weak.
- Ln thiolates are precursors to Ln sulfides.
- Preparation is *via* the 'silylamide route', metathesis or oxidative addition to Ln.

Dithiocarbamates, dithiophosphinates and dithiophosphates

Complexes of the f-block elements with dithiocarbamates have been known since 1908, when thorium, uranyl and neodymium complexes with these ligands were first described. Since then many complexes with the related dithiophosphinate ($R_2PS_2^-$) and dithiophosphate ($(RO)_2PS_2^-$) ligands have been reported. The vast majority of these complexes are prepared by simple metathesis reactions of f-block salts with alkali metal salts of the appropriate dithioacid as shown in Scheme 3.17.

The most common type of lanthanide dithiocarbamate complex is the eight-coordinate $[Ln(S_2CNR_2)_4]^-$ which can have either a distorted square antiprismatic or a distorted dodecahedral coordination geometry. Neutral complexes require the addition of a neutral donor ligand as in $[Ln(S_2CNR_2)_3(bipy)]$. The coordination chemistry of the dithiophosphates and dithiophosphinates is very similar to that of the dithiocarbamates. In the case of $[Y\{S_2P(OEt)_2\}_4]^-$ a P-Y coupling of 5.1 Hz is observed in the ^{31}P NMR spectrum, and has been interpreted by Pinkerton and Earl (1979) as evidence for some covalency in the Y-S bond.

The chemistry of UO_2^{2+} with dithioacid ligands has been particularly well studied because of the potential of these complexes in nuclear fuel reprocessing, and early

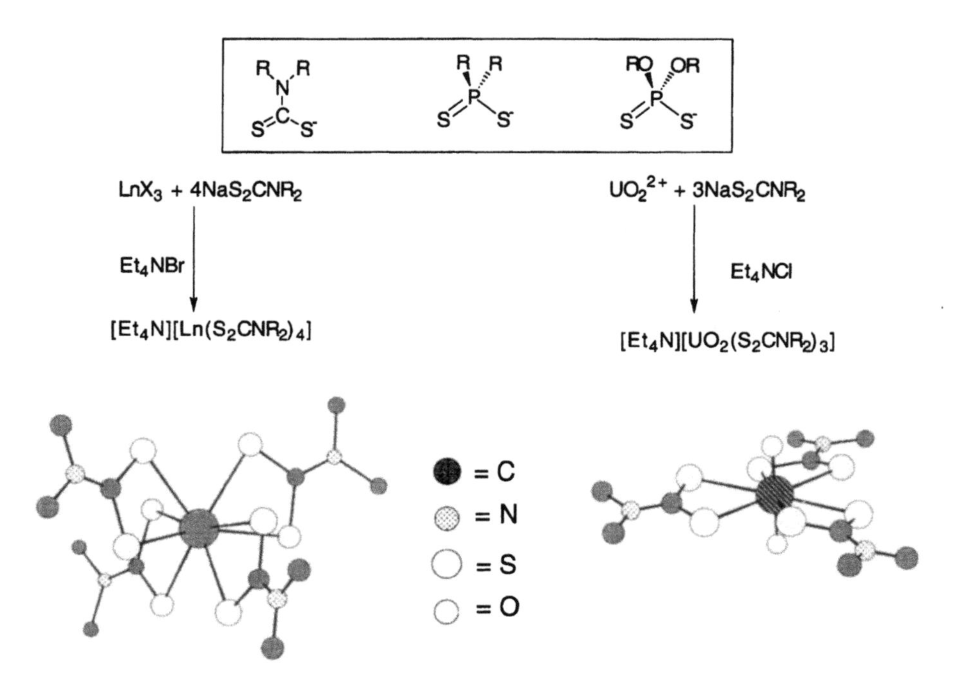

SCHEME 3.17 DITHIOACID LIGANDS AND THE SYNTHESIS AND STRUCTURE OF A LANTHANIDE AND ACTINIDE DITHIOCARBAMATE.

work on actinide complexes with chelating ligands containing sulfur has been reviewed by Castellato, Vidali and Vigato (1979). Anionic eight-coordinate complexes of the type $[UO_2(S_2CNR_2)_3]^-$ are formed by addition of 3 equiv of $S_2CNR_2^-$ to solutions of uranyl salts. Neutral complexes $[UO_2(S_2CNR_2)_2L_2]$ can be isolated in the presence of neutral donors such as Ph_3PO. Uranium (IV) dithiocarbamates can be prepared by insertion of CS_2 into the U-N bonds of $[U(NR_2)_4]$ as shown previously in Scheme 3.8.

3.7 SOLVENT EXTRACTION PROCESSES IN LANTHANIDE AND ACTINIDE SEPARATIONS

Lanthanides occur naturally in two important ores: bastnasite, a fluorocarbonate, and monazite which is a phosphate. Both of these ores contain all of the lanthanides and yttrium, although bastnasite contains only trace quantities of the heavy lanthanides, and monazite contains approximately 10% thorium in addition. Many of the important applications of the lanthanides require separation of the elements, and separation of a series of elements whose chemistry is so similar is not a trivial task. Although separation of naturally occurring thorium and uranium is straightforward (they occur in different ores), mixtures of actinides are produced in nuclear reactors, and separation of transuranics is necessary in nuclear fuel reprocessing. Solvent extraction processes are used in the industrial scale separation of the lanthanides and in the reprocessing of nuclear fuels.

3.7.1 Separation of the lanthanides

The early experiments to isolate the lanthanides used fractional crystallizations which exploited small differences in solubilities of salts. Clearly this is not viable on a large scale and alternative methods had to be developed in order to obtain useful quantities of the separate lanthanides. Fortunately the highly efficient and environmentally friendly process of solvent extraction is now available, and industrial scale lanthanide separations are all performed using this technology.

The solvent extraction process relies on small differences in stability constants of complexes as the series is traversed. An aqueous solution of mixed lanthanide salt (usually nitrate) is extracted, in a process known as 'loading', with an organic solution of a complexing agent (*e.g.* di(2-ethylhexyl)phosphonic acid in odourless kerosene) as shown in Figure 3.29. An extraction coefficient E for a particular Ln^{3+} can be defined as the concentration ratio $[Ln^{3+}]_{organic}/[Ln^{3+}]_{aqueous}$, and because of the differences in stability constants along the lanthanide series, the extraction factor will differ between adjacent elements. The separation factor for two elements A and B is defined as $SF = E_A/E_B$. The larger the value of SF, the fewer cycles of extraction will be required. The extraction of various Ln^{3+} ions with 2-ethylhexyl 2-ethylhexylphosphonic acid at various pH's is shown in Figure 3.29.

$$\underset{aq}{Ln^{3+}} + \underset{org}{3HL} \rightleftharpoons \underset{org}{LnL_3} + \underset{aq}{3H^+}$$

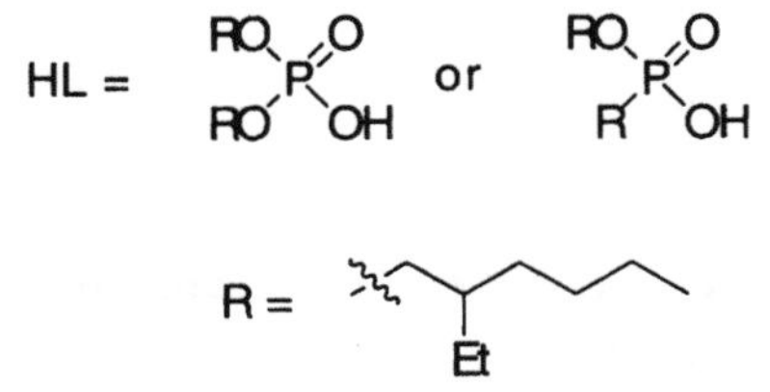

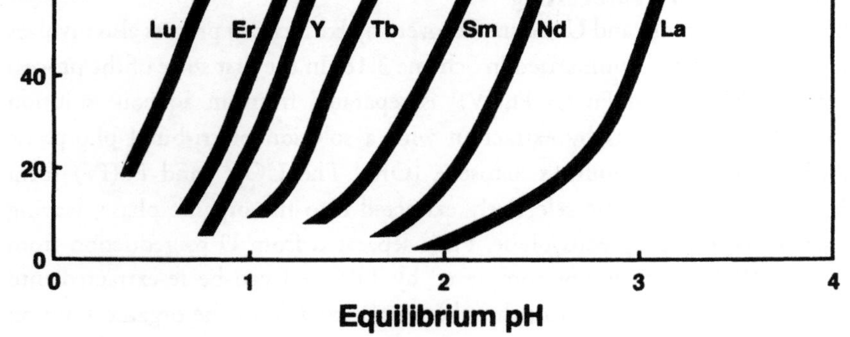

HL = $(RO)_2P(O)OH$ or $(RO)RP(O)OH$

R = 2-ethylhexyl (Et)

FIGURE 3.29 EXTRACTION OF Ln^{3+} IONS WITH 2-ETHYLHEXYL 2-ETHYLHEXYLPHOSPHONIC ACID AT VARIOUS PH'S. (ALBRIGHT AND WILSON AMERICAS).

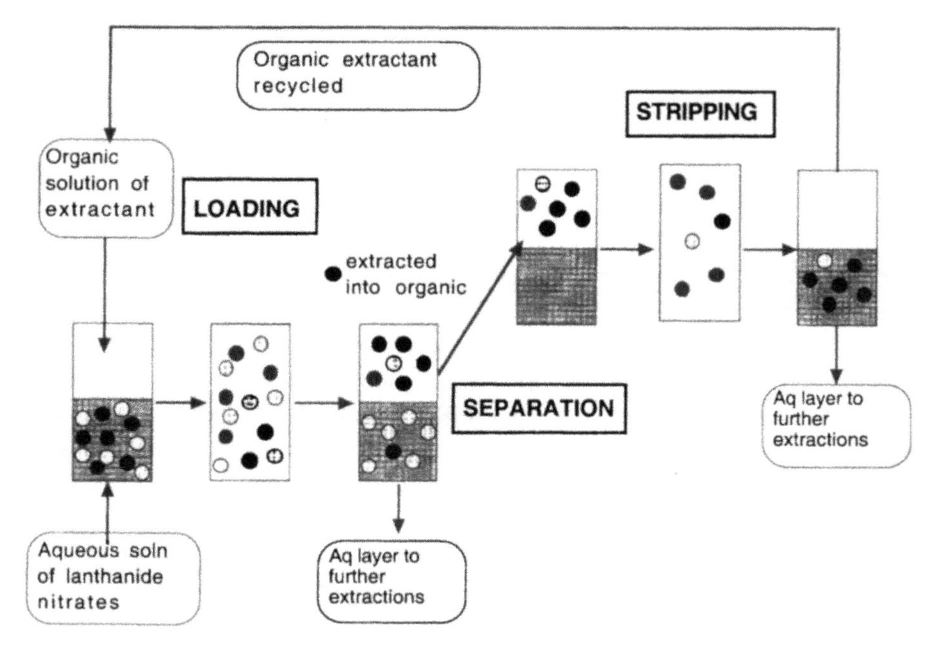

FIGURE 3.30 A SIMPLIFIED DIAGRAM OF THE SEPARATION OF LN^{3+} BY SOLVENT EXTRACTION.

The process is carried out in an industrial version of a separating funnel known as a mixer-settler. After mixing, the organic solution is separated from the aqueous layer, and then 'stripped' into aqueous acid as summarized schematically in Figure 3.30. The enrichment after one extraction cycle is relatively small, but after many cycles, solutions can be obtained which contain single lanthanide salts in very high purity. In practice, for lanthanide separations up to 1500 stages are linked in a countercurrent circuit, where organic and aqueous solutions are cycled around in opposite directions. Individual Ln^{3+} are precipitated from solution as oxalates or hydroxides, which are calcined to give pure oxides.

3.7.2 Nuclear Fuel Reprocessing

The PUREX (Plutonium and Uranium Recovery by Extraction) process also involves solvent extraction and is summarized in Scheme 3.18. In the first stage of the process uranium (as UO_2^{2+}) and Pu (as Pu(IV)) is separated from an aqueous solution containing fission products by extraction with a solution of tri-butyl phosphate ($(BuO)_3P{=}O$, TBP) in odourless kerosene (OK). The UO_2^{2+} and Pu(IV) form complexes with TBP and are selectively extracted into the organic phase, leaving fission products in the aqueous phase. Pu is separated from U by reduction from Pu(IV) to Pu(III), which is not complexed by TBP and can be re-extracted into aqueous solution. Pure UO_2^{2+} can then be re-extracted from the organic solution into very dilute nitric acid.

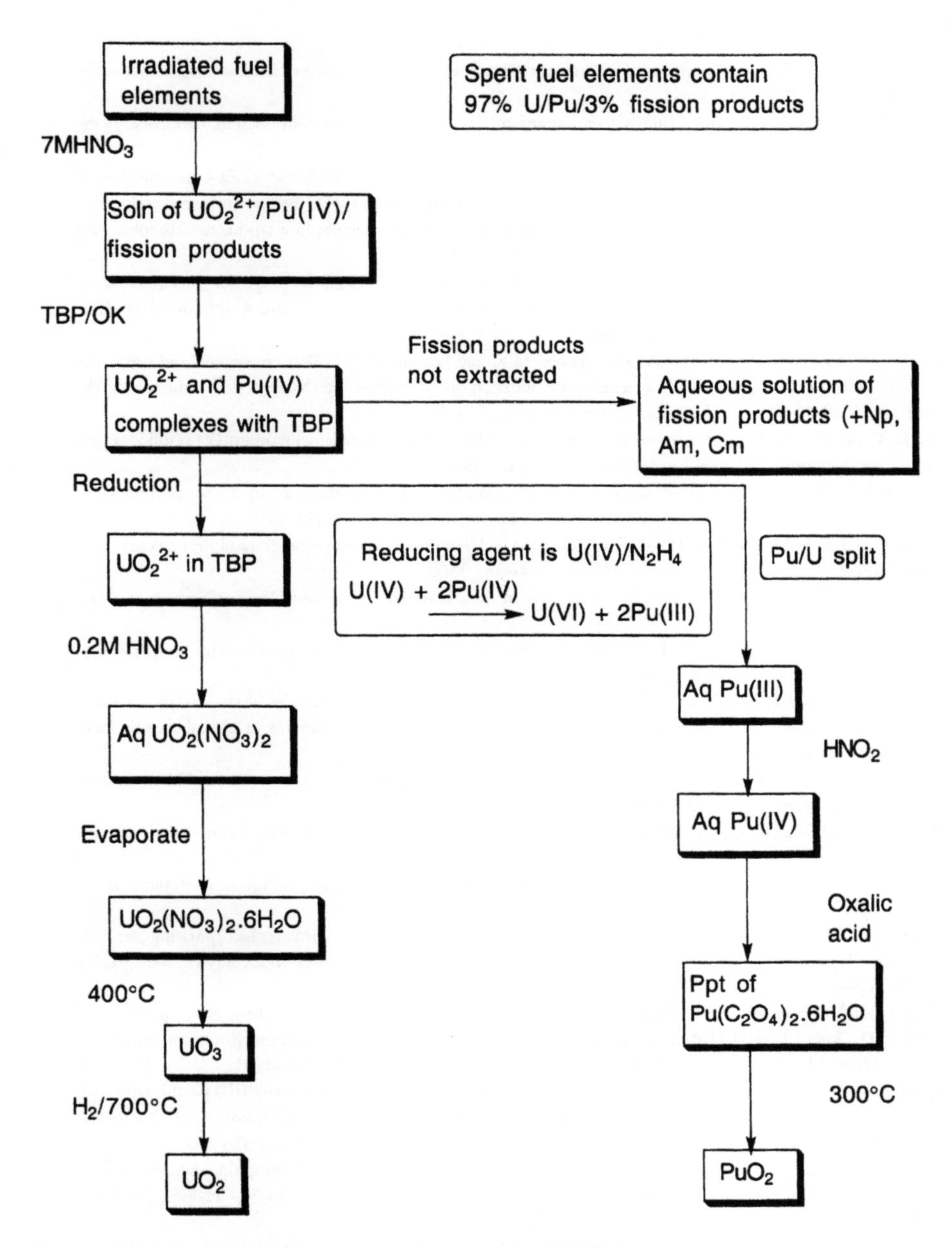

SCHEME 3.18 SCHEMATIC SUMMARY OF THE PUREX PROCESS.

- Solvent extraction is used for the industrial separation of the lanthanides.
- The process depends on differences in stability constants for complexes of adjacent elements.
- $(BuO)_3P{=}O$ coordinates strongly to UO_2^{2+} and is used in the PUREX process for reprocessing nuclear fuel.

REFERENCES

Alexander, V. (1995) Design and synthesis of macrocyclic ligands and their complexes of lanthanides and actinides. *Chem. Rev.*, **95**, 273–342.

Andersen, R.A. (1979) Tris(hexamethyldisilyl)amidouranium(III). Preparation and coordination chemistry. *Inorg. Chem.*, **18**, 1507.

Anwander, R. (1996) Routes to monomeric lanthanide alkoxides. In *Topics in Current Chemistry 179. Organolanthanoid chemistry: synthesis, structure, catalysis*, edited by W.A. Herrmann, pp 149–245. Berlin: Springer-Verlag.

Aspinall, H.C., Moore, S.R. and Smith, A.K. (1992) Synthesis and crystal structure of a lanthanum complex with a diphenylphosphino ligand. *J. Chem. Soc. Dalton Trans.*, 153–156.

Avens, L.R., Bott, S.G., Clark, D.L., Sattelberger, A.P., Watkin, J.G. and Zwick, B.D. (1994) A convenient entry into trivalent actinide chemistry — synthesis and characterization of $AnI_3(THF)_4$ and $An[N(SiMe_3)_2]_3$ (An = U, Np, Pu). *Inorg. Chem.*, **33**, 2248–2256.

Backer-Dirks, J.D.J., Gray, C.J., Hart, F.A., Hursthouse, M.B. and Schoop, B.C. (1979) Preparation and properties of complexes of lanthanides with a hexadentate nitrogen-donor macrocycle: X-ray crystal structure of the complex $[La(NO_3)_3L]$. *J. Chem. Soc., Chem. Commun.*, 774–775.

Bagnall, K.W. and Xing-fu, L. (1982) Some oxygen-donor complexes of cyclopentadienyluranium(IV) N-thiocyanate; steric considerations and stability. *J. Chem. Soc., Dalton Trans.*, 1365–1367.

Barthélemy, P.P., Desreux, J.F. and Massaux, J. (1986) Complexation of lanthanides by linear polyethers in propylenecarbonate: a 'crown-like' behaviour. *J. Chem. Soc., Dalton Trans.*, 2497–2499.

Berthet, J.C. and Ephritikhine, M. (1993) New Reactions of Uranium Amide Complexes Leading to Pentavalent and Cationic Derivatives. *J. Chem. Soc. Chem. Commun.*, 1566.

Berthet, J.C. and Ephritikhine, M. (1998) New Advances in the Chemistry of Uranium Amide Complexes. *Coord. Rev.*, **178–180**, 83–116.

Bradley, D.C., Ghotra, J.S. and Hart, F.A. (1974) Tris-[bis-trimethylsilylamideo]-monochloro-Thorium(IV); A 4-Coordinate Thorium Compound. *Inorg. Nucl. Chem. Lett.*, **10**, 209.

Brennan, J.G., Stults, S.D., Andersen, R.A. and Zalkin, A. (1988) Crystal-structures of $(MeC_5H_4)_3M(L)$ (M = uranium or cerium – L = quinuclidine or $P(OCH_2)_3CEt$) – evidence for uranium to phosphorus pi-back-bonding. *Organometallics*, 7, 1329–1334.

Bünzli, J-C.G. and Pilloud, F. (1989) Macrocyclic effect in lanthanoid complexes with 12-, 15-, 18-, and 21-membered crown ethers. *Inorg. Chem.*, **28**, 2638–2642.

Bünzli, J-C.G. and Wessner, D. (1984) Rare earth complexes with neutral macrocyclic ligands. *Coord. Chem. Reviews*, **60**, 191–253.

Burns, J.H. and Baes, C.F. Jr. (1981) Stability quotients of some lanthanide cryptates in aqueous solutions. *Inorg. Chem.*, **20**, 616–619.

Cary, D.R., Ball, G.E. and Arnold, J. (1995) Trivalent Lanthanide Selenolates and Tellurolates Incorporating Sterically Hindered Ligands and Their Characterization by Multinuclear NMR-Spectroscopy and X-Ray Crystallography. *J. Am. Chem. Soc.*, **117**, 3492–3501.

Casellato, U. and Vigato, P.A. (1978) Actinide complexes with carboxylic acids. *Coord. Chem. Rev.*, **26**, 85–159.

Chandler, C.D., Roger, C. and Hampden-Smith, M.J. (1993) Chemical Aspects of Solution Routes to Perovskite-Phase Mixed-Metal Oxides from Metal-Organic Precursors. *Chem. Rev.*, **93**, 1205–1241.

Ciampolini, M., Mealli, C. and Nardi, N. (1980) Synthesis and properties of some lanthanoid(III) perchlorates with macrocyclic polythioethers of the [18]-crown-6 type. Crystal structure of aquadiperchlorato(1,4,10,13-tetraoxa-7,16-dithiacyclo-octadecane)-lanthanum(III) perchlorate. *J. Chem. Soc. Dalton Trans.*, 376–382.

Coles, S.J., Danopoulos, A.A., Edwards, P.G., Hursthouse, M.B. and Read, P.W. (1995) Diphosphinoamido complexes of thorium(IV) and uranium(IV) and uranium(V)–crystal-structures of $[(ThCl_2[N(CH_2CH_2PPr^i_2)_2]_2)_2]$ and $[(UCl_2[N(CH_2CH_2PEt_2)_2]_2)_2]$. *J. Chem. Soc. Dalton Trans.*, 3401–3408.

Constable, E.C., Chotalia, R. and Tocher, D.A. (1992) The first example of a mono-helical complex of 2,2'/6',2"/6",2'"/6'",2""/6"",2'""-sexipyridine – preparation, crystal and molecular-structure of bis(nitrato-O,O') (2,2'/6',2"/6",2'"/6'",2""/6"",2'"" – sexipyridine)europium(iii)nitrate. *J. Chem. Soc. Chem. Commun.*, 771–773.

Denning, R.G. (1992) Electronic-structure and bonding in actinyl ions. *Structure And Bonding*, **79**, 215–276.

Edwards, P.G., Andersen, R.A. and Zalkin, A. (1981) Tertiary phosphine derivatives of the f-block metals – preparation of $X_4M(Me_2PCH_2CH_2PMe_2)_2$, where X is halide, methyl, or phenoxo and M is thorium or uranium–crystal-structure of tetra(phenoxo)bis[bis(1,2-dimethylphosphino)ethane]-uranium(IV). *J. Am. Chem. Soc.*, **103**, 7792–7794.

Edwards, P.G., Parry, J.S. and Read, P.W. (1995) Homoleptic actinide complexes with chelating diphosphinophosphido ligands – new mode of reactivity with carbon-monoxide and the x-ray crystal-structure of $U(P(CH_2CH_2PMe_2)_2)_2$. *Organometallics*, **14**, 3649–3658.

Eisenstraut, K.J. and Sievers, R.E. (1965) Volatile Rare Earth Chelates. *J. Am. Chem. Soc.*, **87**, 5254–5256.

Evans, W.J., Anwander, R. and Ziller, J.W. (1995) Heteropolyagostic Interactions In Lanthanide(III) Diisopropylamido Complexes. *Inorg. Chem.*, **34**, 5927

Evans, W.J., Drummond, D.K. and Zhang, H. (1988) Synthesis and X-ray Crystal Structure of the Divalent [Bis(trimethylsilyl)amido]samarium Complexes $[(Me_3Si)_2N]_2Sm(THF)_2$ and $\{[(Me_3Si)_2N]Sm(\mu\text{-}I)(DME)(THF)\}_2$. *Inorg. Chem.*, **27**, 575.

Fryzuk, M.D., Haddad, T.S. and Berg, D.J. (1990) Complexes of groups 3, 4, the lanthanides and the actinides containing neutral phosphorus donor ligands. *Coord. Chem. Rev.*, **99**, 137–212.

Fryzuk, M.D., Haddad, T.S. and Rettig, S.J. (1992) Mono(amido-diphosphine) complexes of yttrium – synthesis and X-ray crystal-structure of $(Y(C_3H_5)[N(SiMe_2CH_2PMe_2)_2])_2(\mu\text{-}Cl)_2$. *Organometallics*, **11**, 2967–2969

Ghotra, J.S., Hursthouse, M.B. and Welch, A.J. (1973) Three-co-ordinate Scandium(III) and Europium(III); Crystal and Molecular Structures of their Trishexamethyldisilylamides *J. Chem. Soc. Chem. Commun.*, 669.

Guerriero, P., Tamburini, S. and Vigato, P.A. (1995) From mononuclear to polynuclear macrocyclic or macroacyclic complexes. *Coord. Chem. Reviews*, **139**, 17–243.

Helmholz, L. (1939) The crystal structure of neodymium bromate enneahydrate, $Nd(BrO_3)_3.9H_2O$. *J. Am. Chem. Soc.*, **61**, 1544–1550.

Hitchcock, P.B., Lappert, M.F. and MacKinnon, I.A. (1988) Use of a highly hindered phosphino-alkoxide ligand in the formation of monomeric homoleptic lanthanoid metal-complexes — x-ray structures of $[y(OCBu^t_2CH_2PMe_2)_3]$ $[Nd(OCBu^t_2CH_2PMe_2)_3]$ *J. Chem. Soc. Chem. Commun.*, 1557–155.

Hubert-Pfalzgraf, L.G. (1995) Towards Molecular Design of Homometallic and Heterometallic Precursors for Lanthanide Based Oxide Materials. *New Journal of Chemistry*, **19**, 727-750.

Kahwa, I.A., Folkes, S., Williams, D.J., Ley, S.V., O'Mahoney, C.A. and McPherson, G.L. (1989) The first crystal and molecular structure of lanthanide homodinuclear macrocyclic complexes showing metal-metal pair interactions. *J. Chem. Soc., Chem. Commun.* 1531–1533.

Kumta, P.N. and Risbud, S.H. (1994) Rare-earth chalcogenides – an emerging class of optical-materials. *J. Mat. Sci.* **29**, 1135–1158.

Loncin, M.F., Desreux, J.F. and Merciny, E. (1986) Coordination of lanthanides by two polyamino polycarboxylic macrocycles: formation of highly stable lanthanide complexes. *Inorg. Chem.*, **25**, 2646–2648.

Malandrino, G., Licata, R., Castelli, F., Fragalà, I.L. and Benelli, C. (1995) New Thermally Stable and Highly Volatile Precursors for Lanthanum MOCVD: Synthesis and Characterization of Lanthanum β-Diketonate Glyme Complexes. *Inorg. Chem.*, **34**, 6233–6234.

Mehrotra, R.C. and Singh, A. (1997) Recent Trends in Metal Alkoxide Chemistry. *Prog. in Inorg. Chem.*, **46**, 239-454.

Mehrotra, R.C., Singh, A. and Tripathi, U.M. (1991) Recent advances in Alkoxo and Aryloxo Chemistry of Scandium, Yttrium, and Lanthanoids. *Chem. Rev.*, **91**, 1287–1303.

Miller, S.S., DeFord, D., Marks, T.J. and Weitz, E. (1979) Infrared Photochemistry of a Volatile Uranium Compound with a 10-m Absorption. *J. Am. Chem. Soc.*, **101**, 1036.

Nief, F. (1998) Complexes containing bonds between group 3, lanthanide or actinide metals and non-first-row main group elements (excluding halogens). *Coord. Chem. Rev.*, **168-180**, 13–81.

Nief, F., Riant, P., Ricard, L., Desmurs, P. and Baudry–Barbier, D. (1999) Synthesis and reactivity of bis(phosphoolyl)neodymium(III) and -samarium(III) chlorides and alkyl derivatives. *Eur. J. Inorg. Chem.*, 1041–1045.

Nolan, S.P., Stern, D. and Marks, T.J. (1989) Organo-f-element thermochemistry — absolute metal-ligand bond disruption enthalpies in bis(pentamethylcyclopentadienyl)samarium hydrocarbyl, hydride, dialkylamide, alkoxide, halide, thiolate, and phosphide complexes — implications for organolanthanide bonding and reactivity. *J. Am. Chem. Soc.*, **111**, 7844–7853.

Ouchi, A., Suzuki, Y., Ohki, Y. and Koizumi, Y. Structure of rare earth carboxylates in dimeric and polymeric forms. *Coord. Chem. Rev.*, **92**, 29–44.

Piguet, C., Bünzli, J.-C.G., Bernardinelli, G., Hopfgartner, G. and Williams, A.F. (1993) Self assembly and photophysical properties of lanthanide dinuclear triple helical complexes. *J. Am. Chem. Soc.*, **115**, 8197–8206.

Pinkerton, A.A. and Earl, W.L. (1979) Nuclear spin-spin coupling to ^{89}Y. A model to investigate the question of covalent bonding in lanthanide dithiophosphinate complexes. *J. Chem. Soc. Dalton Trans.*, 1347–1349.

Rabe, G.W., Riede, J. and Schier, A. (1996a) Synthesis, X-ray crystal structure determination, and NMR spectroscopic investigation of two homoleptic four-coordinate lanthanide complexes: Trivalent $(^tBu_2P)_2La[(\mu\text{-}P^tBu_2)_2Li(thf)]$ and divalent $Yb[(\mu\text{-}P^tBu_2)_2Li(thf)]_2$. *Inorg. Chem.*, **35**, 40–45.

Rabe, G.W., Riede, J. and Schier, A. (1996b) Structural diversity in lanthanide phosphido complexes. Formation and X-ray crystal structure determination of an unusual dinuclear phosphido complex .of divalent samarium: $[(Me_3Si)_2P]Sm[\mu\text{-}P(SiMe_3)_2]_3Sm(thf)_3.C_7H_8$. *Organometallics*, **15**, 439–441.

Rabe, G.W., Yap, G.P.A. and Rheingold, A.L. (1995) Divalent lanthanide chemistry — 3 synthetic routes to samarium(II) and ytterbium(II) bis(phosphido) species including the structural characterization of $Yb[PPh_2]_2(THF)_4$ and $Sm[PPh_2]_2(N\text{-}MeIm)_4$. *Inorg. Chem.*, **34**, 4521–4522.

Reynolds, J.G., Zalkin, A. and Templeton, D.H. (1976) Crystal Structure and Optical and Magnetic Properties of Tetrakis(diethylamido)uranium(IV), a Five-Coordinate Dimeric Complex in the Solid State. *Inorg. Chem.*, **15**, 2498.

Sabbatini, N., Guardigli, M. and Lehn, J-M. (1993) Luminescent lanthanide complexes as photochemical supramolecular devices. *Coord. Chem. Rev.*, **123**, 201–228.

Schumann, H., Palamidis, E., Schmid, G. and Boese, R. (1986) $[(Cp_2Lu(\mu\text{-}PPh_2)_2Li(tmeda)].1/2C_6H_5CH_3$, the first organolanthanoid-phosphane compound to be characterized by X-ray structure analysis. *Angew. Chem. Int. Ed. Engl.*, **25**, 718–719.

Sweeting, L.M. and Rheingold, A.L. (1987) Crystal Disorder and Triboluminescence: Triethylammonium Tetrakis(dibenzoylmethanato)europate. *J. Am. Chem. Soc.*, **109**, 2652–2658.

Tilley, T.D., Andersen, R.A. and Zalkin, A. (1982) Tertiary Phosphine Complexes of the f-Block Metals. Crystal Structure of $Yb\{N(SiMe_3)_2\}_2(Me_2PCH_2CH_2PMe_2)$: Evidence for Ytterbium-γ-Carbon Interaction. *J. Am. Chem. Soc.*, **104**, 3725.

Van den Hende, J.R., Hitchcock, P.B. and Lappert, M.F. (1994) 3-coordinate neutral ligand-free ytterbium(II) complexes $[(YbX(\mu\text{-}X))_2]$ (X=OAr **1** or $OCBu^t{}_3$**3**) or $[(Yb(NR_2)(m\text{-}X))_2]$ (X=$OCBut_3$ **2** or OAr **4**) (Ar=$C_6H_2Bu^t{}_2$-2,6-Me-4, R=$SiMe_3$) — the x-ray structure of **1** and **2**. *J. Chem. Soc. Chem. Commun.*, 1413–1414.

Van der Sluys, W.G. and Sattelberger, A.P. (1990) Actinide Alkoxide Chemistry. *Chem. Rev.*, **90**, 1027–1040.

CHAPTER 4

ORGANOMETALLIC CHEMISTRY

There are a relatively small number of homoleptic σ-bonded organo-lanthanide and -actinide complexes. Without exception they are highly sensitive to oxygen and moisture; in some cases they are also thermally unstable. The high and usually non-specific reactivity of these compounds makes their chemistry difficult to study, and the chemistry of the Ln- or An-to-C bond has mainly been elucidated in heteroleptic compounds, usually with π-bonded cyclopentadienyl supporting (also known as spectator or ancillary) ligands. This chapter describes the rapidly-developing area of σ- and π-bonded organo-lanthanide and -actinide chemistry. It will open with a relatively brief description of homoleptic σ-bonded organometallics; the larger part of the chapter will deal with complexes containing π-bonded ligands, mainly cyclopentadienyls and substituted derivatives.

4.1 HOMOLEPTIC σ-BONDED ORGANOMETALLICS

Most of the chemistry of the Ln/An-C σ-bond has been investigated in heteroleptic complexes with cyclopentadienyl spectator ligands. There are, however, some well-defined homoleptic σ-bonded organo lanthanides and actinides, which are the subject of this section. The large size of lanthanide and actinide ions means that stable complexes are only obtained by the use of chelating or bulky ligands, or the formation of anionic 'ate' complexes with higher coordination numbers. The bonding is primarily ionic in nature and so these complexes are highly sensitive to oxygen and moisture and their geometry is defined only by steric factors.

4.1.1 Synthetic routes

Most σ-bonded organo-lanthanide and -actinide complexes are prepared by metathesis reactions of either an anhydrous metal chloride or a metal aryloxide with an organolithium reagent. The alkoxide route is often preferable as retention of alkali metal halides in the product may occur with both lanthanides and actinides.

4.1.2 Lanthanide complexes

The great majority of well-characterized homoleptic σ-bonded organolanthanides are in oxidation state 3. The first σ-bonded organolanthanide was the tetrahedral 'ate' complex $[Lu(Ar)_4]^-$; analogous complexes for the larger lanthanides were not available, however use of the more sterically demanding Bu^t ligand allowed preparation of $[LnBu^t_4]^-$ complexes for Ln= Sm, Er, Yb. Thermally stable hexamethyl 'ate' complexes $[Li(tmeda)]_3[LnMe_6]$ were prepared by Schumann *et al.* (1984) for yttrium and for all the lanthanides except Eu, which is reduced by the excess MeLi. The first neutral homoleptic complex was reported by Hitchcock *et al.* (1988) and used the bulky $CH(SiMe_3)_2$ ligand to give the pyramidal three-coordinate $[Ln\{CH(SiMe_3)_2\}_3]$ for Ln = La to Lu. These complexes were prepared from $[Ln(O\text{-}2,6\text{-}Bu^t_2\text{-}C_6H_3)_3]$ and $LiCH(SiMe_3)_2$ as metathesis with $LnCl_3$ resulted in retention of LiCl in the product. Chelating *ortho*-((dimethylamino)methyl) phenyl was used by Wayda, Atwood and Hunter (1984) to give 6-coordinate $[Ln(2\text{-}((Me_2N)CH_2)C_6H_4)_3]$ for Ln = Er, Yb and Lu, but analogous complexes were not available for earlier lanthanides. Alkyl complexes of Yb(II) and Eu(II) have been reported by Eaborn *et al.* (1994 and 1996). Synthetic routes to several homoleptic lanthanide alkyl complexes are summarized in Scheme 4.1.

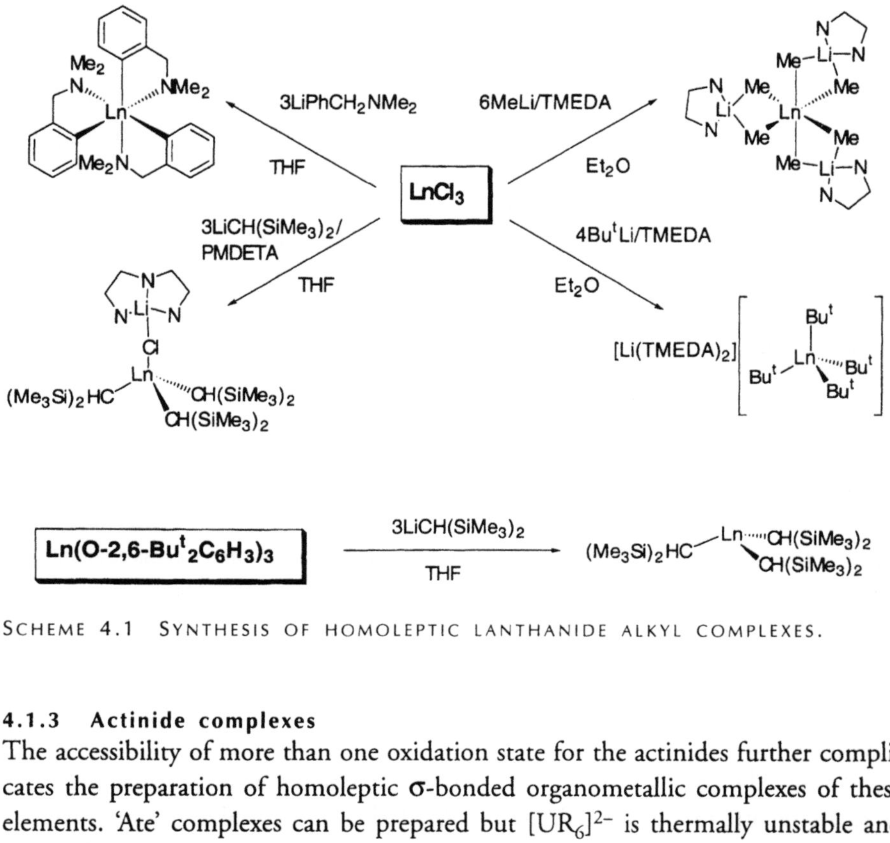

SCHEME 4.1 SYNTHESIS OF HOMOLEPTIC LANTHANIDE ALKYL COMPLEXES.

4.1.3 Actinide complexes

The accessibility of more than one oxidation state for the actinides further complicates the preparation of homoleptic σ-bonded organometallic complexes of these elements. 'Ate' complexes can be prepared but $[UR_6]^{2-}$ is thermally unstable and decomposes by β-hydride elimination. The addition of excess LiR in an attempt to prepare $[UR_7]^{3-}$ leads to reduction of U(IV) to U(III). Th(IV) is less susceptible to reduction than U(IV) and the seven-coordinate complex $[Li(tmeda)]_3[ThMe_7]$ can be prepared; its structure is shown in Figure 4.1. The neutral four-coordinate complex $[ThAr_4]$ (Ar = 2, 5-Me_2-$C_6H_3CH_2$) is also known. Soon after the preparation of the first neutral homoleptic tri-alkyl lanthanide complexes, Van der Sluys, Burns and Sattelberger (1989) prepared the first actinide analogue, $[U(CH(SiMe_3)_2)_3]$ which, as expected, has a pyramidal structure like those of the analogous Ln complexes.

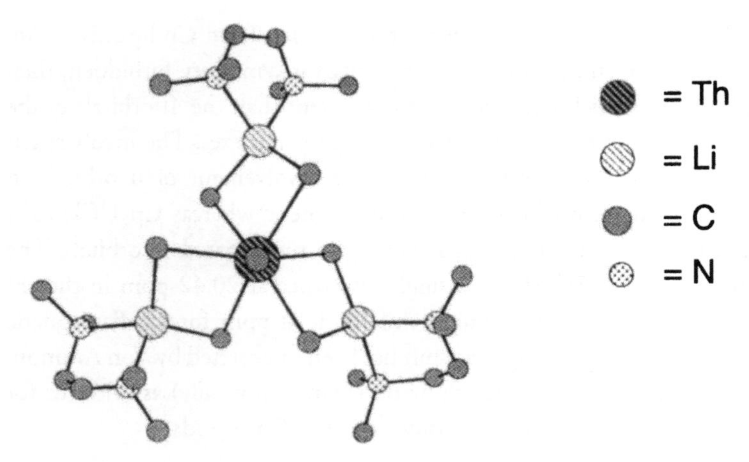

FIGURE 4.1 STRUCTURE OF $[Li(TMEDA)]_3[ThMe_7]$ VIEWED ALONG THE Me-Th BOND.

Virtually all the reactivity studies of lanthanide- and actinide-C σ-bonds have been carried out on heteroleptic complexes, usually with cyclopentadienyl supporting ligands. Discussion of reactivity will therefore be deferred until Section 4.3.

- Synthesis of homoleptic alkyl compounds of the lanthanides and actinides is by metathesis of the anhydrous halides or alkoxides with organo-Li reagents under strictly anaerobic conditions.
- Coordinative saturation is achieved by formation of anionic 'ate' complexes with high coordination numbers or by use of bulky or chelating ligands.
- All σ-bonded organometallics of the f-elements are extremely air and moisture sensitive.

4.2 CYCLOPENTADIENYL COMPLEXES

Cyclopentadienyl, Cp, and its substituted derivatives, are almost ubiquitous as supporting ligands in organo-f-element chemistry. Although susceptible to protonolysis, the lower basicity of Cp^- compared with R^- means that it is much less reactive than alkyl ligands in heteroleptic complexes. The large Ln and An ions require bulky or multidentate ligands in order to satisfy their coordination requirements, and a range of substituents can be introduced into the Cp ring in order to achieve this. Bulky substituted cyclopentadienyls can stabilize otherwise highly reactive alkyl ligands, and the introduction of chiral substituents onto Cp rings has opened up a whole area of enantioselective organo-lanthanide catalysts which is described in Chapter 5.

There have been a considerable number of theoretical and spectroscopic investigations of bonding in lanthanide and actinide cyclopentadienyl complexes. See for example Bursten and Strittmatter (1991), DiBella *et al.* (1996), DiBella *et al.* (1994). There is no real evidence for significant covalent contributions to the bonding in cyclopentadienyl lanthanide complexes. The situation is somewhat different in organoactinide chemistry, and there is evidence for donation from Cp ligands to An, usually into the 6d orbitals. If donation into 6d orbitals is symmetry forbidden, then donation into 5f orbitals, which have greater spatial extent than the 4f orbitals of the lanthanides, may occur as in the C_{3v} symmetric Cp_3An complexes. The involvement of 5f orbitals in bonding is very different from the involvement of d orbitals in transition metal complexes: Cp_2MoCl_2 ($4d^2$) is diamagnetic whereas Cp_2UCl_2 ($5f^2$) is paramagnetic with the unpaired electrons residing in metal-based 5f orbitals. The tetrahedral complex [Cp_4U] ($5f^2$) shows a single resonance at 20.42 ppm in the ^1H NMR spectrum, compared with a chemical shift of 1.10 ppm for its diamagnetic analogue [Cp_4Th]. This large paramagnetic shift has been interpreted by Von Ammon, Kanellakopulos and Fischer (1970) (though by no means universally) as evidence for significant delocalization of 5f electron density onto the Cp ligands.

4.2.1 Lanthanide Cyclopentadienyl Complexes

In this section the synthesis, structures and reactivity of Ln(III) complexes of the type Cp_3Ln and Cp_2LnX (X=halide, alkyl or hydride) will be described, as well as some Ln(II) complexes Cp_2Ln. Schumann, Meese-Marktscheffel and Esser (1995) have reviewed lanthanide(III) chemistry with cyclopentadienyl and other π-donor ligands.

Lanthanide tris(cyclopentadienyl) complexes

The first organometallic complexes of the lanthanides were the *tris*(cyclopentadienyls) [$LnCp_3$] reported by Wilkinson and Birmingham (1954), four years after the first synthesis of ferrocene. These complexes are prepared by the reaction of NaCp with anhydrous $LnCl_3$ in THF; the initial product, [$LnCp_3$(THF)] remains in solution and is easily separated from the insoluble NaCl by-product. As the bonding between Cp and Ln is ionic in nature, $LnCp_3$ are very moisture sensitive and must be prepared and handled under strictly anhydrous conditions. [$LnCp_3$(THF)] may be desolvated by heating *in vacuo* and the resulting $LnCp_3$ show a range of structures dictated by the need to satisfy the coordination requirements of the large Ln^{3+} ion while minimizing repulsive interactions between ligands.

The requirements of large La^{3+} ion are satisfied by formation of a polymer where each La atom is coordinated to three η^5 Cp ligands and has a η^2 interaction with one other Cp ligand. In the intermediate ionic radius range (including Sm) a simple monomeric species is formed. The smallest lanthanide, Lu, is unable to accommodate three η^5 Cp ligands, and $LuCp_3$ adopts a polymeric structure with each Lu coordinated to two η^5 Cp ligands and two μ_2-Cp ligands. Structures of $LnCp_3$ species are illustrated in Figure 4.2 and the synthesis is as shown below:

$$LnCl_3 + 3NaCp \xrightarrow{THF} 3NaCl_{ppt} + [Cp_3Ln(THF)] \xrightarrow{200^\circ C,\ vacuum} [Cp_3Ln]$$

The $LnCp_3$ unit is a Lewis acid and several adducts of the type $LnCp_3L$ have been characterized, for example where L = PPh_3, RNC, or even a transition metal carbonyl complex. In the latter case, an oxygen lone pair of the CO group acts as a Lewis

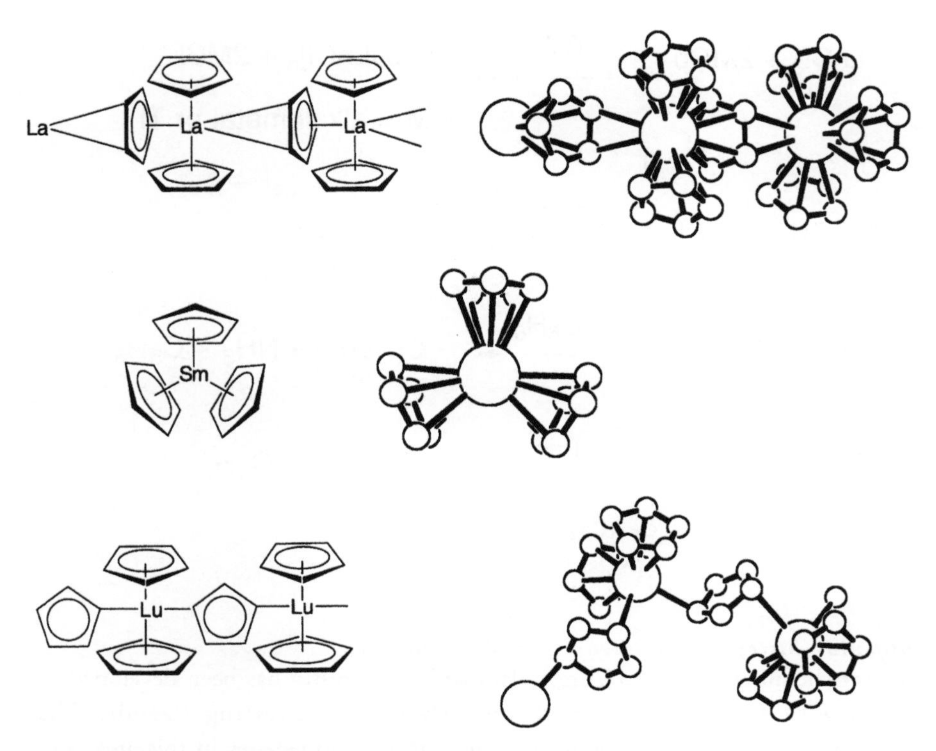

FIGURE 4.2 STRUCTURES OF $[Cp_3Ln]$ COMPLEXES.

base towards the Ln atom. Isocyanide is known as a π-acceptor as well as a σ-donor in transition metal chemistry; on coordination to Cp_3Ln, n(C≡N) generally increases by 60–70 cm^{-1}, signifying that there is σ-donation to Ln but essentially no Ln-CNR π backbonding. The crystal structure of $[Cp_3Pr(CNC_6H_{11})$ is shown in Figure 4.3.

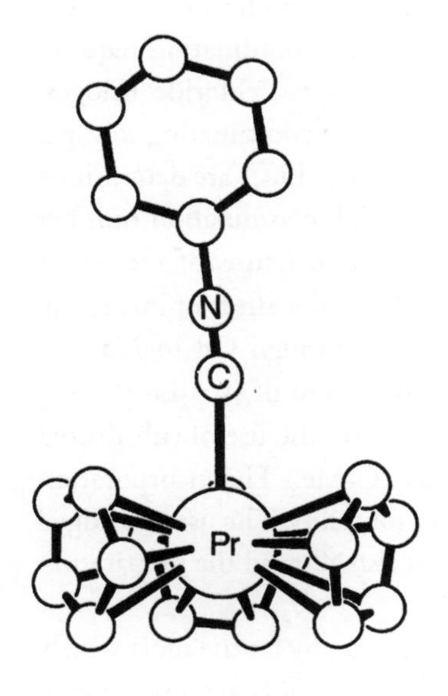

FIGURE 4.3 STRUCTURE OF $[Cp_3Pr(CNC_6H_{11})]$.

$$LnCl_3 + 2MCp \xrightarrow{THF} [Cp_2LnCl]_n + 2MCl$$

M = alkali metal or Tl

$$2LnCp_3 + LnCl_3 \xrightarrow{THF} 3Cp_2LnCl$$

$$Cp_3Ln + NH_4Cl \xrightarrow{C_6H_6} Cp_2LnCl + NH_3 + C_5H_6$$

$$Cp_3Ln + HCl \xrightarrow{THF} Cp_2LnCl + C_5H_6$$

SCHEME 4.2 SYNTHESIS OF [CP_2LNCL] COMPLEXES.

Lanthanide bis(cyclopentadienyl) halide complexes

The overwhelming majority of organolanthanide chemistry has been developed in systems with Cp (or substituted derivatives) as supporting ligands. The *bis*(cyclopentadienyl) lanthanide halides are crucial intermediates in this area, and a great deal of effort has been devoted to their synthesis and structural characterization. Synthetic routes are outlined in Scheme 4.2. Metathesis reactions between MCp and $LnCl_3$ may appear to give the most straightforward route, but can lead to formation of heterometallic 'ate' complexes $[Cp_2Ln(\mu\text{-}Cl)_2M]$, from which elimination of the alkali metal can prove difficult. The synthesis by ligand redistribution reaction between $LnCl_3$ and two equivalents of $LnCp_3$ is another useful route and illustrates the lability of many lanthanide complexes and the stability of Cp_2LnCl.

The unsolvated Cp_2LnCl fragment is Lewis acidic and the coordination requirements of Ln are satisfied by dimer or oligomer formation *via* chloride bridges. Solvates, sometimes monomeric, are formed in the presence of coordinating solvents such as THF. As with Cp_3Ln compounds, the structures of Cp_2LnCl are determined by a combination of the steric requirements of Ln for a high coordination number and the requirement to minimize interligand repulsions. The structures of a selection of Cp_2LnCl complexes are shown in Figure 4.4, which shows the effect of increasing Ln^{3+} radius from the smaller late lanthanides (Er to Yb) through Gd to Nd.

The simple C_5H_5 ligand is too small to allow the formation of $[Cp_2Ln(\mu\text{-}Cl)]_2$ for Ln larger than Y; the formation of simple dimers requires the use of substituted Cp ligands such as $(Me_3Si)_2C_5H_3$ or the even more bulky C_5Me_5. The incorporation of substituents in the Cp ring also leads to an increase in solubility. The use of bridged cyclopentadienyl ligands as described in Section 4.2 has also allowed the preparation of Cp_2LnCl type complexes for early Ln.

Although the *bis*(cyclopentadienyl) lanthanide chlorides are by far the most widely studied Cp_2LnX compounds, analogues with X=F, Br and CF_3SO_3 are also known.

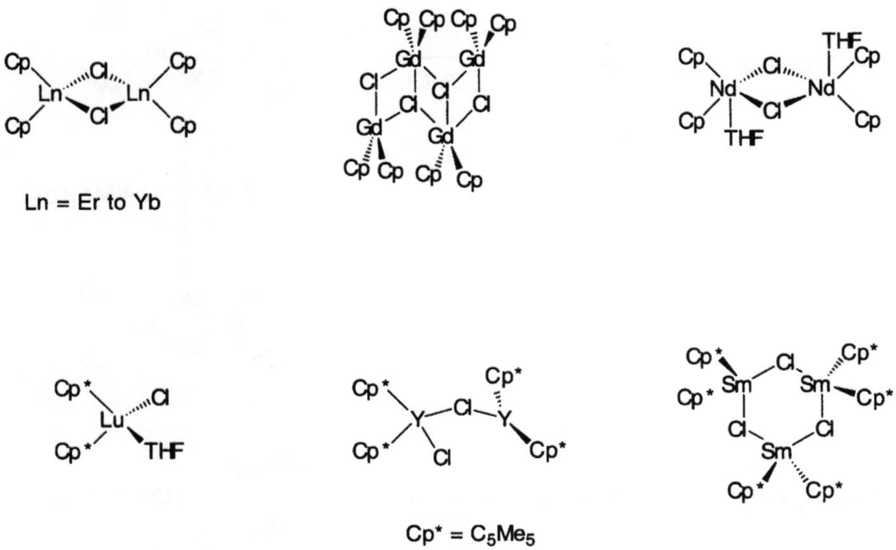

Ln = Er to Yb

$Cp^* = C_5Me_5$

FIGURE 4.4 STRUCTURES OF $[Cp_2LnCl]$ COMPLEXES.

Bis(cyclopentadienyl) alkyl lanthanide complexes

The most important synthetic route to these compounds is metathesis of Cp_2LnCl complexes with RLi. For Me compounds, an alternative route is available *via* the cleavage of the heterometallic complexes $[Cp_2Ln(\mu\text{-}Me)_2AlMe_2]$ with a Lewis base such as pyridine as shown in Scheme 4.3. The highly reactive bis(pentamethyl-cylcopentadienyl) complexes of Sm(II) and Yb(II) described later readily undergo 1-electron oxidation reactions, and transmetallation with R_2Hg gives $Cp^*{}_2LnR$ and metallic Hg.

The structures adopted by bis(cyclopentadienyl) alkyl lanthanide complexes are mainly dependent on steric factors: the size of Ln, the bulk of R and the degree of substitution on the cyclopentadienyl ligands. For example $[Cp_2YMe]_2$ is a dimeric molecule with two μ_2 bridging Me groups whereas $[(C_5Me_5)_2YCH(SiMe_3)_2]$, with bulky C_5Me_5 ligands and a large alkyl group, is monomeric. $[(C_5Me_5)_2LuMe]_2$ shows an unusual structure in the solid state with one terminal Me ligand and one Me ligand forming a linear bridge between the Lu atoms. In solution this complex exists in a monomer-dimer equilibrium. Reactions of Cp_2LnR complexes will be described in Section 4.3.

Bis(cyclopentadienyl) lanthanide hydrides

The first organolanthanide complexes containing Ln-H bonds were tetrahydroborates $[Cp_2LnBH_4(THF)]$ prepared by Marks and Grynkewich in 1976. Lanthanide hydrides of the type $[Cp_2LnH]$ were reported by Evans, Engerer and Coleson (1981), Evans *et al.* (1982) and by Schumann and Genthe (1981). The bis(cyclopentadienyl) lanthanide hydrides may be formed by β-hydride elimination from lanthanide alkyls, by metathesis of $[Cp_2LnCl]$ with MH, or by hydrogenolysis of the Ln-R bond, which

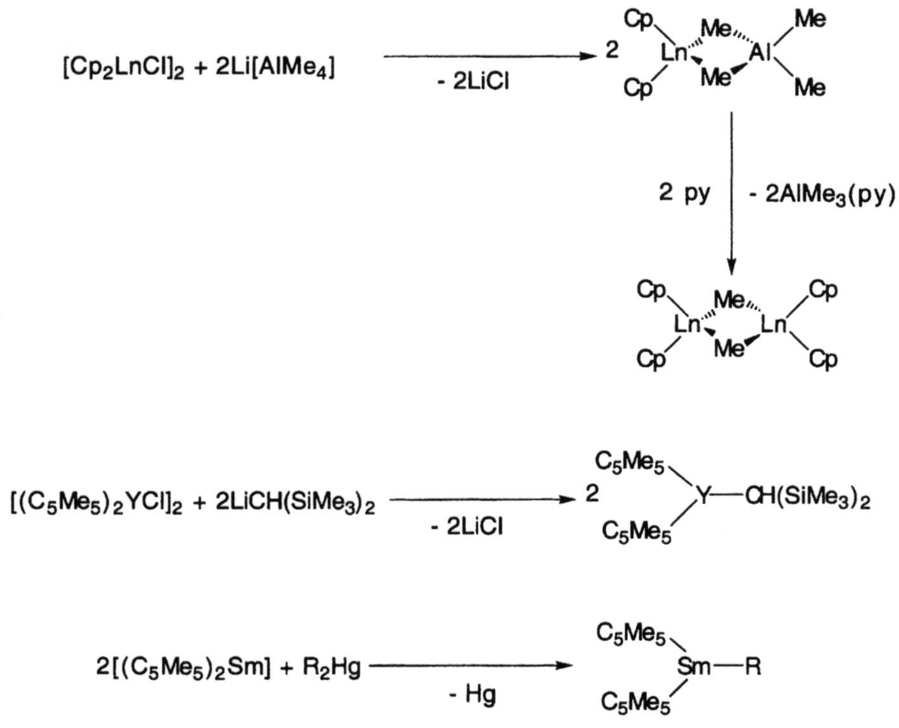

$$[Cp_2LnCl]_2 + 2Li[AlMe_4] \xrightarrow{-2LiCl} 2\ Cp_2Ln(\mu\text{-}Me)_2AlMe_2$$

$$2\ Cp_2Ln(\mu\text{-}Me)_2AlMe_2 \xrightarrow[-2AlMe_3(py)]{2\ py} Cp_2Ln(\mu\text{-}Me)_2LnCp_2$$

$$[(C_5Me_5)_2YCl]_2 + 2LiCH(SiMe_3)_2 \xrightarrow{-2LiCl} 2\ (C_5Me_5)_2Y{-}CH(SiMe_3)_2$$

$$2[(C_5Me_5)_2Sm] + R_2Hg \xrightarrow{-Hg} (C_5Me_5)_2Sm{-}R$$

SCHEME 4.3 SYNTHESIS OF [Cp_2LnR] COMPLEXES.

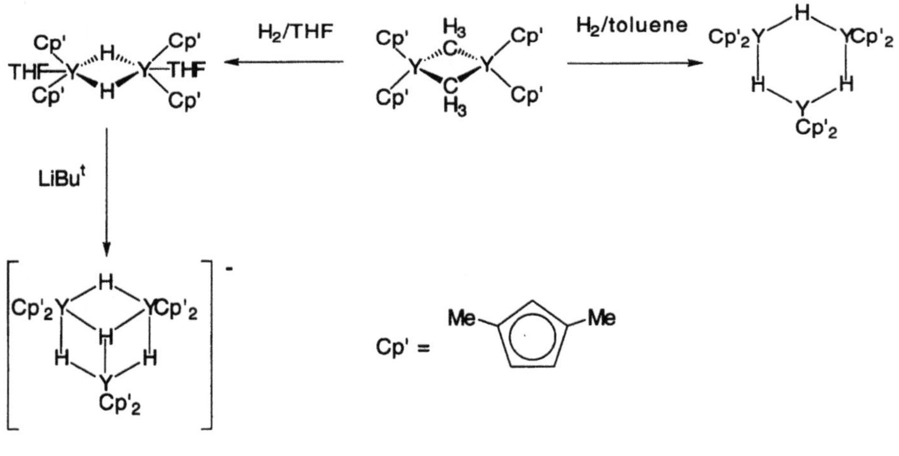

$$[Cp'_2Y(THF)(\mu\text{-}H)]_2 \xleftarrow{H_2/THF} [Cp'_2Y(\mu\text{-}CH_3)]_2 \xrightarrow{H_2/toluene} [Cp'_2Y(\mu\text{-}H)]_3$$

$$[Cp'_2Y(THF)(\mu\text{-}H)]_2 \xrightarrow{LiBu^t} [(Cp'_2Y)_3(\mu\text{-}H)_3(\mu_3\text{-}H)]^-$$

Cp' = Me–C_5H_3–Me

SCHEME 4.4 SYNTHESIS OF BIS(CYCLOPENTADIENYL)YTTRIUM HYDRIDES.

is the preferred synthetic route as shown in Scheme 4.4. As with other cyclopentadienyl lanthanide complexes Cp_2LnH adopt a range of structures determined by the steric requirements of Ln and the repulsive interactions between the ligands. The ^1H NMR chemical shifts of the hydride ligands are rather variable: the μ_2-H of $[(C_5H_4Me)_2Y(THF)(\mu\text{-}H)]_2$ resonates as a triplet ($J_{Y\text{-}H}$ = 27 Hz) at 2.02 ppm, whereas the Lu analogue shows a singlet at 4.69 ppm; $[(C_5Me_5)_2LuH]$ shows a singlet at 9.27 ppm.

Bis(cyclopentadienyl) lanthanide hydrides catalyze a range of important organic transformations including hydrogenationof alkenes, hydroamination of aminoalkenes and aminoalkynes, oligomerization/polymerization of alkenes and alkynes, polymerization of methylmethacrylate. These reactions will be described in detail in Chapter 5.

- $[Cp_2LnX]$ adopt a variety of sructures dependent on the ionic radius of Ln^{3+}.
- $[Cp_2LnCl]$ are important starting materials for the synthesis of organolanthanides.
- $[Cp_2LnR]$ and $[Cp_2LnH]$ catalyze several important organic transformations.

Bis(cyclopentadienyl)lanthanide(II) complexes

Oxidation state +2 is important for Sm, Eu and Yb, and the first compounds of the type Cp_2Ln were obtained for Eu and Yb by reaction of liquid NH_3 solutions of the metals with cyclopentadiene. The products were initially obtained as the solvates $[Cp_2Ln(NH_3)_2]$ which can be desolvated by heating to 120–200°C *in vacuo*. This route is not applicable to Sm complexes as Sm is insoluble in liquid NH_3. Ln^{2+} ions are considerably larger than their Ln^{3+} counterparts and therefore higher coordination numbers or bulkier ligands are needed to satisfy their coordination requirements. The first monomeric unsolvated Cp_2Ln complexes were thus prepared using the very bulky C_5Me_5 ligand.

Evans, Hughes and Hanusa (1986) showed that the structures of $[(C_5Me_5)_2Ln]$ have a 'bent' metallocene geometry with a centroid-Ln-centroid angle of approximately 140°, compared with an angle of 136.7° in the corresponding solvates. The crystal structures of solvated and unsolvated complexes are shown in Figure 4.5. One explanation of this anomalous structure is that the bending leads to a polarized molecule for which favourable dipole-dipole interactions with other molecules are possible in the solid state. This dipole-dipole interaction can only be favourable for large metal ions; it is observed for Cp_2Ca but not for Cp_2Mg. A more recent molecular mechanics study by Timofeeva, Lii and Allinger (1995) suggests that crystal packing effects are not the most important contribution to the bending and that intramolecular non-bonded interactions between cyclopentadienyl groups in the complex are the major cause of the bending. Whatever its cause, the bending of the Cp_2Ln fragment is only possible because of the non-directional ionic nature of the Ln-Cp interaction.

The large size of the Ln^{2+} ions means that $[(C_5Me_5)_2Ln]$ complexes are coordinatively unsaturated and adduct formation with Lewis bases such as ethers, N-donors or even P-donors is therefore an important reaction. Coordination of H_2 and ethene to

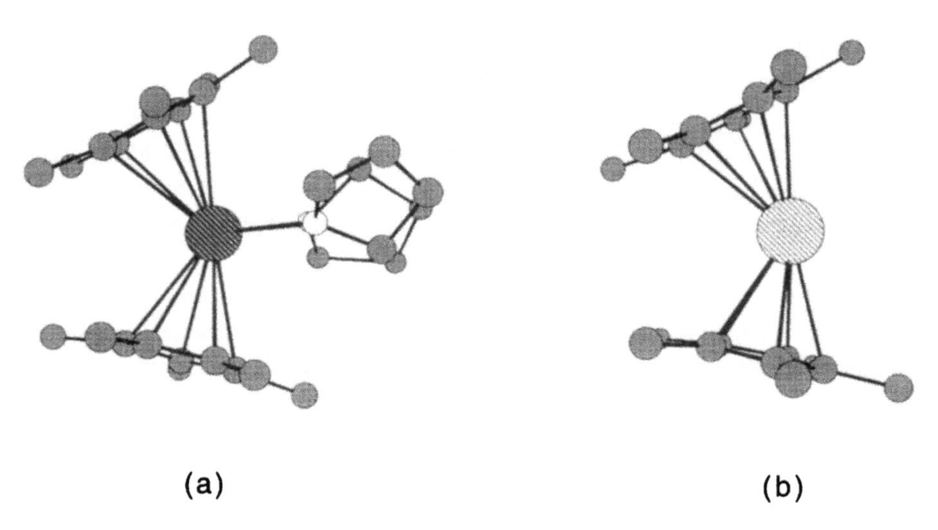

FIGURE 4.5 (A) STRUCTURE OF $[(C_5Me_5)_2Sm(THF)_2]$ (B) STRUCTURE OF $[(C_5Me_5)_2Sm]$.

$[(C_5Me_5)_2Eu]$ has been observed in solution by NMR spectroscopy; as expected the bonding is essentially donation of electron density from the ligand to Eu: there is no evidence for lanthanide-to-ligand back bonding in these complexes. Reaction of $[(C_5Me_5)_2Yb]$ with $[Pt(PPh_3)_2(C_2H_4)]$ resulted in the formation of the first isolable alkene complex of an f-element; it was reasoned that the electron rich nature of ethene coordinated to $Pt(PPh_3)_2$ would favour coordination to Yb(II). An alkyne complex was formed with but-2-yne.

One-electron reduction chemistry is also very important for $[(C_5Me_5)_2Ln]$, especially for Sm where unique reactivity has been observed. Evans, Ullibari and Ziller (1988) have shown that reaction of $[(C_5Me_5)_2Sm]$ with N_2 leads to reversible formation of $[\{(C_5Me_5)_2Sm\}_2(\mu_2\text{-}N_2)]$ where crystallographically determined bond lengths are consistent with a N=N double bond and oxidation to Sm(III). With CO $[(C_5Me_5)_2Sm(THF)]$ gives a ketenecarboxylate where three CO units have been reductively coupled in a unique way. These reactions are summarized in Scheme 4.5.

$[(C_5Me_5)_2Eu]$ is much less susceptible to oxidation than $[(C_5Me_5)_2Sm]$ as demonstrated by reactions of $[(C_5Me_5)_2Ln(OEt_2)]$ withPhCCH (Scheme 4.6): the alkyne acts as an acid towards the Eu complex, resulting in formation of a Eu(II) product, whereas it acts as a oxidizing agent towards the Sm complex giving a Sm(III) product. On addition to $[(C_5Me_5)_2Yb(OEt_2)]$, PhCCH acts as both an acid and a reducing agent and results in formation of a mixed valence Yb(III)/Yb(II) product as reported by Boncella, Tilley and Andersen (1984).

- $[(C_5Me_5)_2Ln]$ are known for Sm, Eu and Yb.
- They adopt a 'bent metallocene' structure.
- Reactivity includes adduct formation with Lewis bases and one-electron oxidation of Ln.

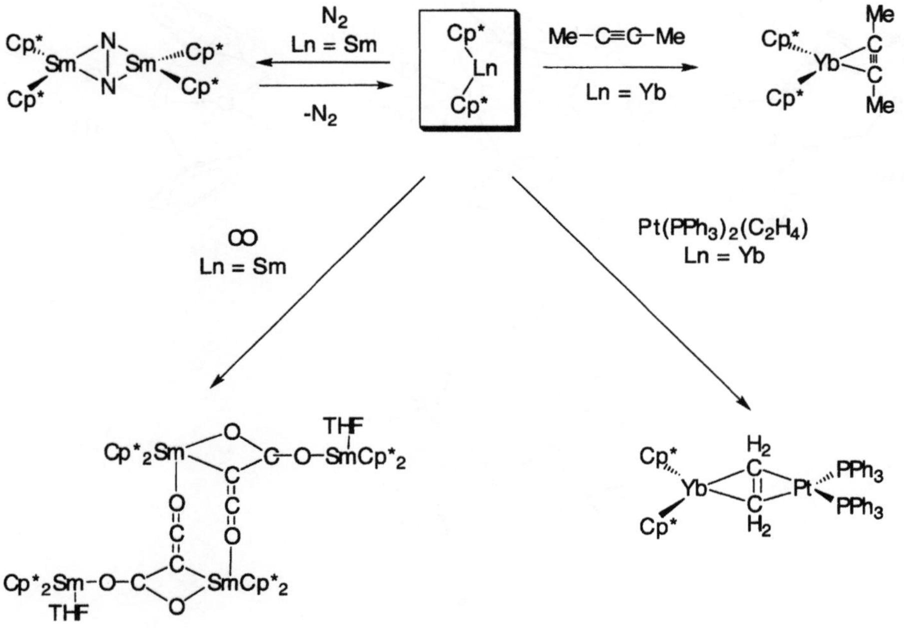

SCHEME 4.5 REACTIONS OF $[(C_5Me_5)_2Ln]$.

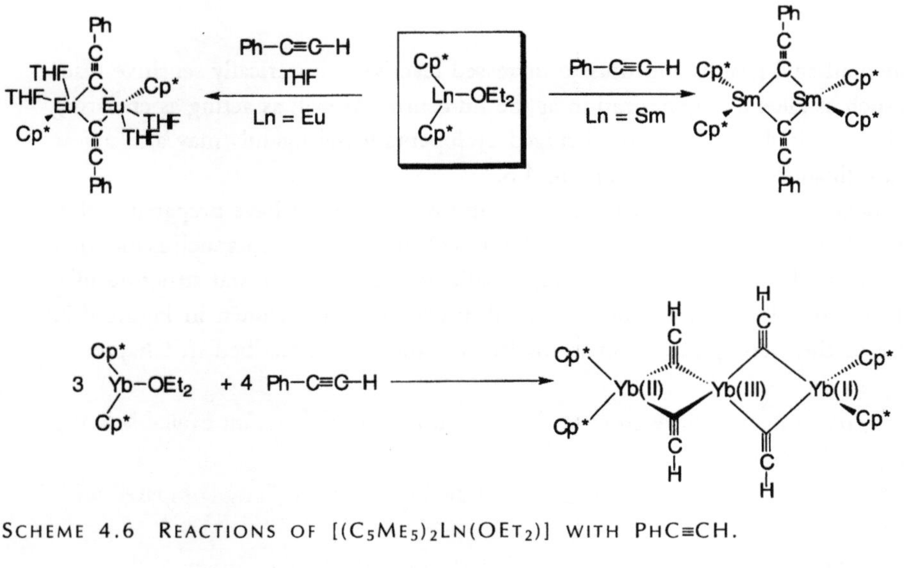

SCHEME 4.6 REACTIONS OF $[(C_5Me_5)_2Ln(OEt_2)]$ WITH PhC≡CH.

Bridged cyclopentadienyl ligands
Two cyclopentadienyl rings can be linked (*e.g.* with a CH_2 bridge) and the resulting chelating ligands can lead to modified reactivity of their complexes. The -$CH_2CH_2CH_2$- or -CH_2OCH_2- linked cyclopentadienyls allow isolation of bis(cyclopentadienyl) chlorides for the large early lanthanides as shown in Figure 4.6. Linking of rings with a one-atom bridge such as $SiMe_2$ results in 'tying back' of the

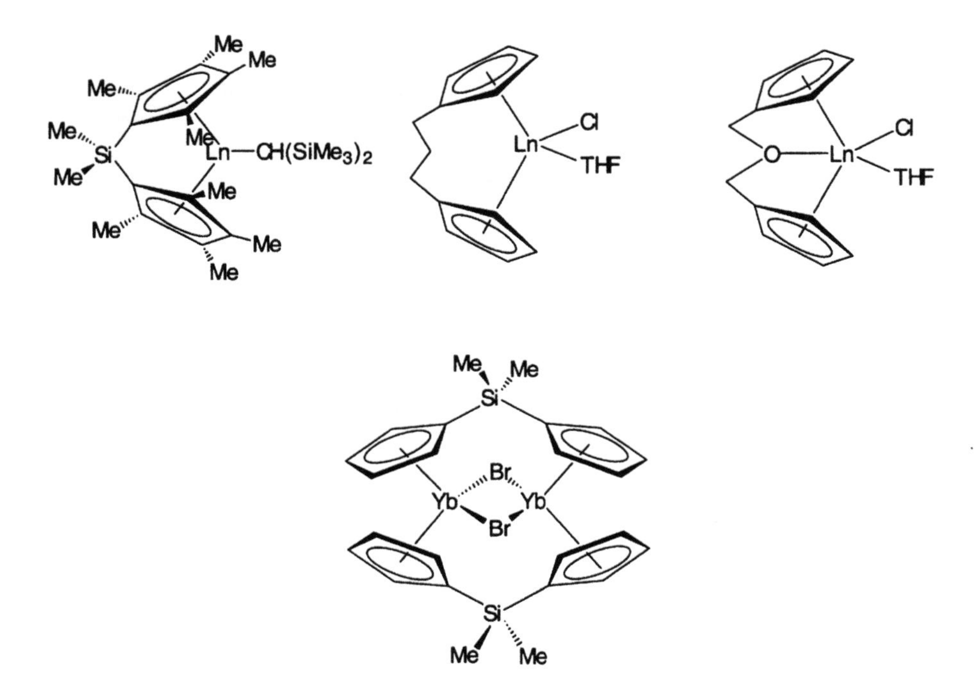

FIGURE 4.6 STRUCTURES OF LN COMPLEXES WITH BRIDGED CYCLOPENTADIENYL LIGANDS.

cyclopentadienyl groups and thus to increased reactivity in sterically sensitive reactions such as alkene polymerization at the Ln centre. As well as acting as chelating ligands to a single metal centre, bridged cyclopentadienyl ligands may also act as bridging ligands, as shown in Figure 4.6.

Giardello *et al.* (1994) and Haar, Stern and Marks (1996) have prepared highly enantioselective chiral catalysts by introduction of chiral substituents such as menthyl or neomenthyl onto one of the cyclopentadienyl rings. The crystal structure of a chiral Sm complex with a neomenthyl substituted ligand is shown in Figure 4.7; the use of these complexes in enantioselective catalysis is described in Chapter 5.

- Bridged Cp ligands have allowed the isolation of complexes not available using simple Cp ligands.
- Complexes with 'tied-back' Cp ligands show increased reactivity compared with their simple Cp analogues.
- Chiral substituents on 'tied-back' Cp ligands result in chiral complexes which are useful in enantioselective catalysis.

4.2.2 Actinide cyclopentadienyl complexes

Actinide(III) tris(cyclopentadienyl) complexes

The +3 oxidation state is not common for Th and U, but Cp_3An have been prepared for both these elements. $[Cp_3U]$ can be prepared by metathesis of UCl_3 with KCp,

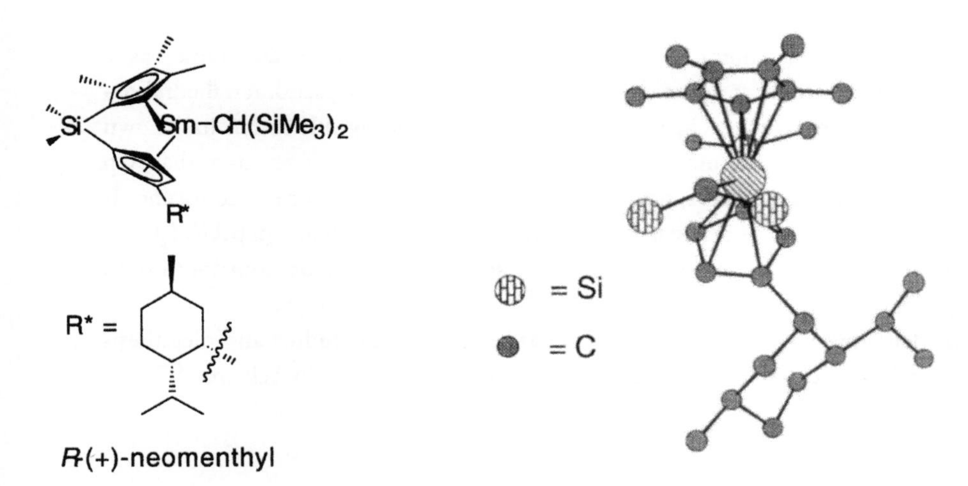

FIGURE 4.7 STRUCTURE OF A CHIRAL ORGANOSAMARIUM COMPLEX. FOR CLARITY ME GROUPS ARE OMITTED FROM $CH(SiMe_3)_2$ LIGAND.

or by reduction of U(IV) complexes *e.g.* reaction of Cp_3UCl with NaH or reaction of metallic U with 3 equiv of $[Cp_4U]$. $[Cp_3Th]$ is prepared by reduction of $[Cp_3ThCl]$ with Na naphthalenide in THF. These compounds are Lewis acidic, like their lanthanide analogues, and readily form adducts with Lewis bases. However Brennan, Andersen and Robbins (1986) found that $[(C_5H_4SiMe_3)_3U]$, unlike $[Cp_3Ln]$, forms an adduct with CO which is stable in solution for prolonged periods of time. This complex shows a $\nu(C{\equiv}O)$ of 1976 cm^{-1} suggesting significant U-to-CO π backbonding *via* a 5f orbital and has been characterized crystallographically by Conejo *et al.* (1999). An analogous isocyanide complex was isolated and characterized crystallographically.

- $[Cp_3An]$ are Lewis acidic.
- There is some evidence for U-to-CO back bonding in $[(C_5H_4SiMe_3)_3U(CO)]$.

Actinide(IV) tetrakis(cyclopentadienyl) complexes

Homoleptic Cp_4An (An = Th, U, Np) are well-known and can be prepared by metathesis of $AnCl_4$ with KCp. Reaction of the U(IV) amide complex $[U(NEt_2)_4]$ with excess cyclopentadiene has also been reported to give Cp_4U. The X-ray structure of Cp_4U has been determined: it is a tetrahedral molecule with four η^5-Cp ligands. The average Cp-to-U distance is slightly longer than in other Cp complexes of U(IV) indicating a high degree of steric crowding at the metal. The chemistry of Cp_4An is very little developed.

Actinide(IV) tris(cyclopentadienyl) halide complexes

In 1956, two years after the first report of an organolanthanide complex, Reynolds and Wilkinson reported $[Cp_3UCl]$, the first organoactinide complex. It was synthesized by the metathesis reaction between UCl_4 and three equivalents of CpNa in

THF. A more convenient preparation uses TlCp, which is air stable, as a cyclopentadienyl transfer agent. [Cp_3UCl] has the expected pseudotetrahedral geometry, with approximate C_{3v} symmetry, and powder X-ray diffraction has shown the Th complex to be isomorphous. [Cp_3UCl] is much less air-sensitive than Cp complexes of the lanthanides, and does not react with $FeCl_2$ to form ferrocene. It can even be dissolved in deoxygenated water to form the aquo ion $[Cp_3U(H_2O)]^+$. These observations suggest that there is a significant covalent contribution to U-Cp bond.

The derivative chemistry of [Cp_3UCl] has been very well studied and metathesis reactions have yielded a wide range of complexes as outlined in Scheme 4.7.

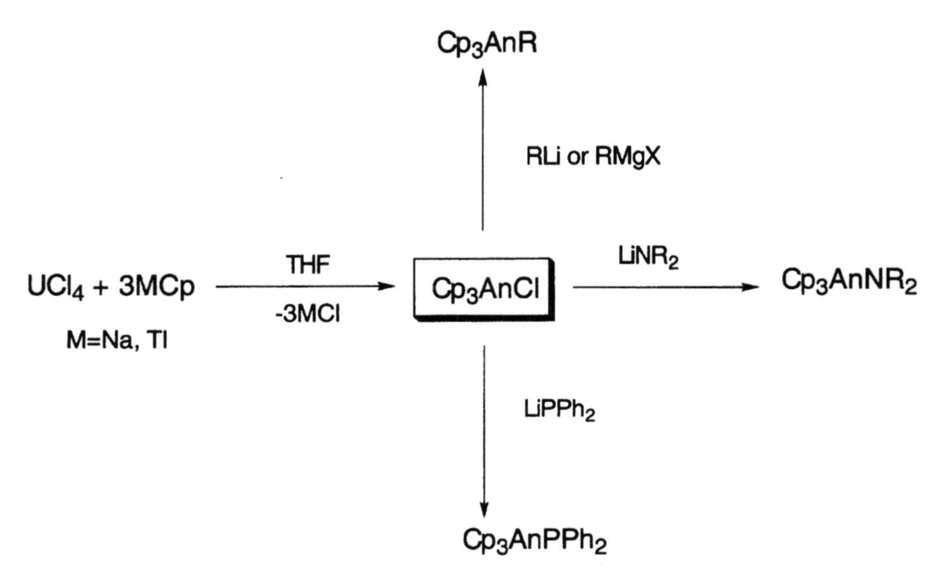

SCHEME 4.7 METATHESIS REACTIONS OF [Cp_3UCl].

Tris(cyclopentadienyl)alkylactinides

Alkyl actinide complexes of the type Cp_3AnR may be prepared by metathesis reactions between Cp_3AnCl and RLi or RMgX, or by reaction of Cp_3AnCl with RX in the presence of Na/Hg. They have been shown by X-ray diffraction to adopt a pseudotetrahedral structure. Their reactivity is discussed in Section 4.3 and some results of thermochemical studies are described in Section 4.5.

Tris(cyclopentadienyl)actinide hydrides

Simple Cp_3AnH complexes cannot be prepared, probably because of facile elimination of H_2. The use of bulkier substituted cyclopentadienyls such as $C_5H_4Bu^t$ prevents the bimolecular H_2 elimination reaction and [$(C_5H_4R)_3UH$] have been prepared by reaction of [$(C_5H_4R)_3UCl$] with K[$BHEt_3$] and fully characterized for R=Bu^t, $SiMe_3$ or PPh_2. Complex hydrides with BH_4^- or AlH_4^- have also been prepared and utilized as starting materials for synthesis of other complexes.

- $[Cp_3AnCl]$ are useful starting materials.
- There is some evidence for covalent contributions to U-Cp bonding in $[Cp_3UCl]$, which is much less reactive than $[Cp_2LnCl]$.
- $[Cp_3AnR]$ are well known.
- $[Cp_3AnH]$ cannot be isolated; bulkier ligands such as $(C_5H_4Bu^t)$ must be used.

Bis(cyclopentadienyl)actinide dihalides

The bis(cyclopentadienyl)actinide dihalides are analogues of the metallocene dichlorides $[Cp_2MCl_2]$ (M=Ti, Zr, Hf) for which a very extensive chemistry has been developed. However, $[Cp_2AnCl_2]$ are kinetically unstable and undergo rapid ligand redistribution reactions to form $[CpAnCl_3]$ and $[Cp_3AnCl]$ species. This ligand redistribution is only prevented by using bulky substituted cyclopentadienyls such as C_5Me_5. Reaction of $AnCl_4$ with two equivalents of the Grignard reagent C_5Me_5MgCl leads to selective formation of the desired product. Use of alkali metal reagents MC_5Me_5 in THF leads to reduction of U(IV) to U(III). A wide range of compounds is available via metathesis reactions from $[(C_5Me_5)_2AnCl_2]$.

Bis(cyclopentadienyl)alkylactinides

Pentamethylcyclopentadienyl is probably the best supporting ligand for the stabilization of alkyl actinide complexes, and compounds of the type $[(C_5Me_5)_2AnR_2]$ have been well studied for both Th and U. These complexes are relatively straightforward to prepare; their synthesis is usually *via* the metathesis of $[(C_5Me_5)_2AnCl_2]$ with the appropriate alkyl lithium. A neutron diffraction study of the neopentyl complex $[(C_5Me_5)_2Th(CH_2Bu^t)_2]$ by Bruno *et al.* (1986) showed this complex (Figure 4.8) to have unequal Th-C (neopentyl) bond lengths of 2.456(4) Å and 2.543(4) Å.

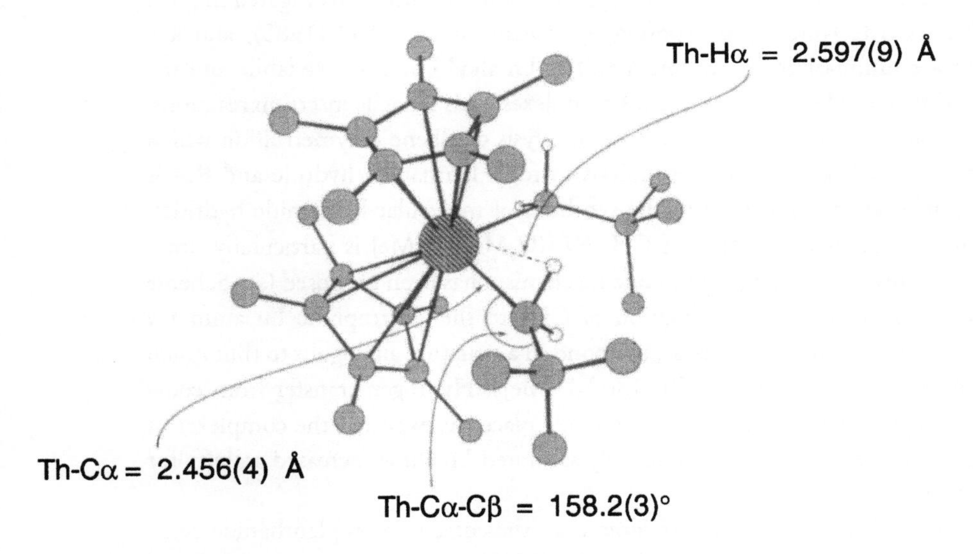

FIGURE 4.8 STRUCTURE OF $[(C_5Me_5)_2Th(CH_2CMe_3)_2]$.

A very open Th-C-C bond angle of 158.2(3) Å and a Th-H distance of 2.597(9) Å were also observed, and together with theoretical studies these structural observations have been explained by agostic interactions between Th and the α-C-H bond. The very facile cyclometallation reactions which occur for these complexes (Sections 4.3 and 4.5) are also consistent with these observations.

4.3 REACTIONS OF σ-BONDED ORGANOLANTHANIDES AND -ACTINIDES

Important reactions characterized for σ-bonded organolanthanides and/or organoactinides are:

- Protonolysis.
- Hydrogenolysis.
- Insertion.
- β-hydride elimination.
- β-alkyl elimination.
- C-H activation.

The Ln-C σ-bond is almost totally ionic in nature and so these compounds are highly carbanionic. They therefore undergo protonolysis, eliminating RH on reaction with even the weakest of acids, and must be handled under rigorously anhydrous conditions. This sensitivity to protic reagents has been exploited in the synthesis of lanthanide complexes with S, P or Se donor atoms, which is described in more detail in Chapter 3. Reactions involving formal changes in oxidation state (*e.g.* oxidative addition/reductive elimination) are unimportant for Ln(III) alkyls. Most of the chemistry of the Ln-C σ-bond has been studied in heteroleptic complexes such as Cp_2LnR, where the chemistry is simplified by the presence of relatively unreactive spectator ligands. The chemistry of the Ln-alkyl bond has been investigated in great detail for $[(C_5Me_5)_2LuMe]$ as reported by Watson and Parshall (1985), and key reactions are summarized in Scheme 4.8. The Ln-alkyl bond is very labile and the Lewis acidity of the Ln atom in Cp_2LnR complexes leads to facile insertion reactions. Facile alkene insertion leads to very ready catalysis of alkene polymerization which is described in Chapter 5. Hydrogenolysis to form lanthanide hydride and RH is a well established reaction used in the synthesis of molecular lanthanide hydrides.

The unprecedented activation of CH_4 by $[(C_5Me_5)_2LuMe]$ is particularly noteworthy. This reaction is complex but one mechanism has been proposed (see Scheme 4.9) which involves weak coordination of CH_4 to the electrophilic Lu atom *via* donation of electron density from a C-H bond in a structure analogous to (but much less stable than) that of dimeric $[(C_5Me_5)_2LuMe]_2$. Hydrogen transfer from coordinated CH_4 to the CH_3 ligand can then take place. As expected the complexes of the larger and therefore less coordinatively saturated Ln show increased activity for CH_4 activation.

Actinide alkyls are somewhat less ionic than the corresponding lanthanide complexes (*e.g.* a hybrid $6d(z^2)$-$7p(z)$ orbital in Cp_3An^+ has appropriate energy and symmetry to accept electron density from R^-), and the An-C σ-bond is less carbanionic

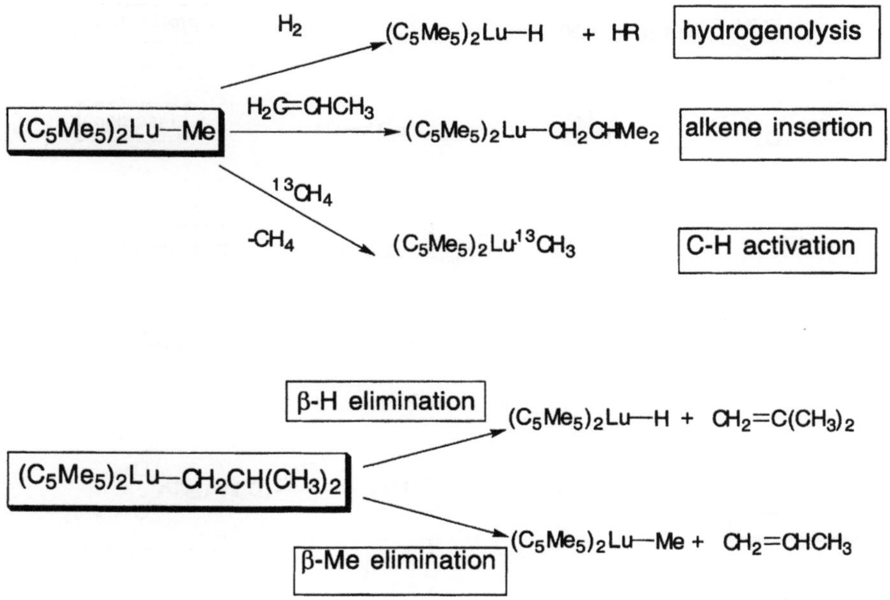

SCHEME 4.8 REACTIONS OF BIS(PENTAMETHYLCYCLOPENTADIENYL)ALKYL LUTETIUM COMPLEXES.

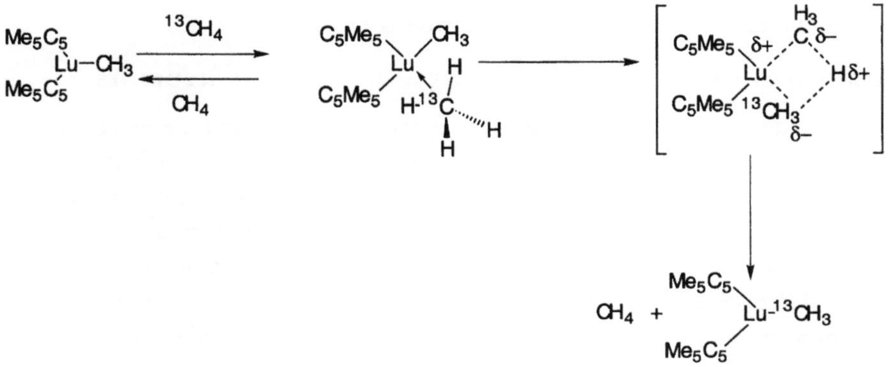

SCHEME 4.9 ACTIVATION OF CH_4 BY $[(C_5ME_5)_2LUME]$.

for U than for Th. Reactivity of the actinide alkyls is greatly affected by the steric demands of the spectator ligands *e.g.* the sterically congested Cp_3AnR compounds are thermally quite stable and do not undergo β-hydride elimination whereas the Cp_2AnR_2 analogues are much more reactive. Protonolysis, hydrogenolysis, CO or isocyanide insertion and C-H activation have all been characterized for An alkyl complexes, and a selection of reactions is shown in Scheme 4.10. Bond disruption energies have been measured by Schock *et al.* (1988) for An-R complexes; it has been found that D(Th-R) is generally greater by approximately 10–30 $kJmol^{-1}$ than the corresponding

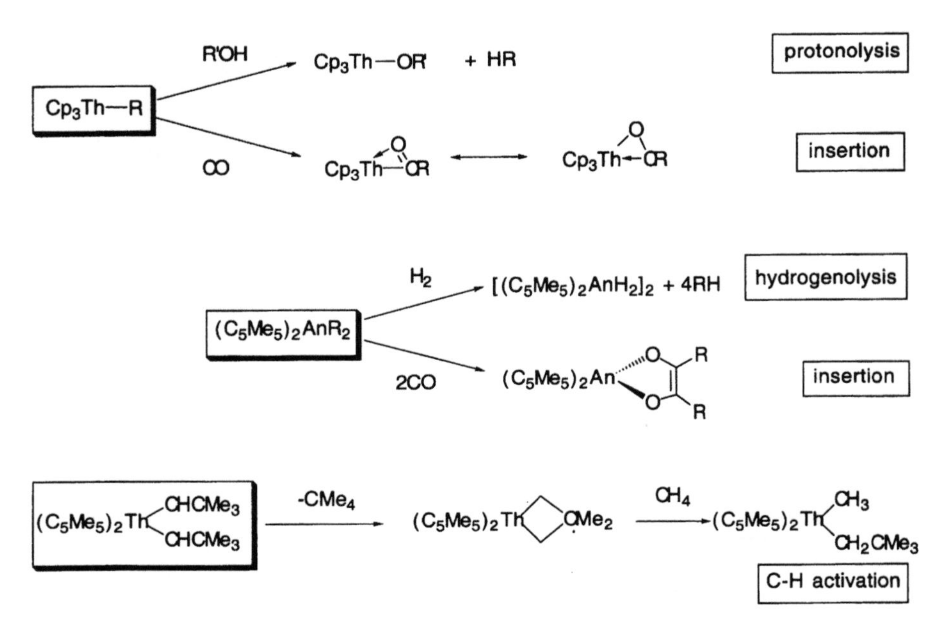

SCHEME 4.10 REACTIONS OF ACTINIDE ALKYL COMPLEXES.

D(U-R), and these energies have been correlated with the relative rates of CO insertion. In the case of An=U, the CO insertion is reversible.

4.4 REACTIONS OF ORGANOLANTHANIDE AND -ACTINIDE HYDRIDES

Organolanthanide and -actinide hydrides are extremely reactive and are excellent catalysts for a number of synthetically useful organic transformations which are described in Chapter 5. Marks and co-workers have made detailed calorimetric measurements of Ln-H and An-H bond disruption enthalpies and used these in rationalizing some of the reactions of these compounds (see Section 4.5). The $[(C_5Me_5)_2Lu\text{-}H]$ system has been studied in great detail by Watson (1983), and C-H activation by this complex has been observed for both sp^2 and sp^3 C-H bonds as summarized in Scheme 4.11.

Insertion of alkenes is an important reaction for both Ln-H and An-H; this is the first step in the highly efficient alkene hydrogenation catalyzed by organolanthanide- and organoactinide hydrides. Examples are given in Chapter 5.

- An/Ln-hydride bonds are highly polar.
- Activation of aromatic and aliphatic C-H bonds has been observed.
- Insertion of alkenes into Ln/An-H is an important catalytic reaction.

4.5 THERMOCHEMICAL STUDIES OF REACTIVITY OF ORGANO-LANTHANIDE AND -ACTINIDE ALKYLS AND HYDRIDES

Values of bond disruption enthalpies D(M-X) for a range of organolanthanide and organoactinide species have been determined by Marks and co-workers; a selection

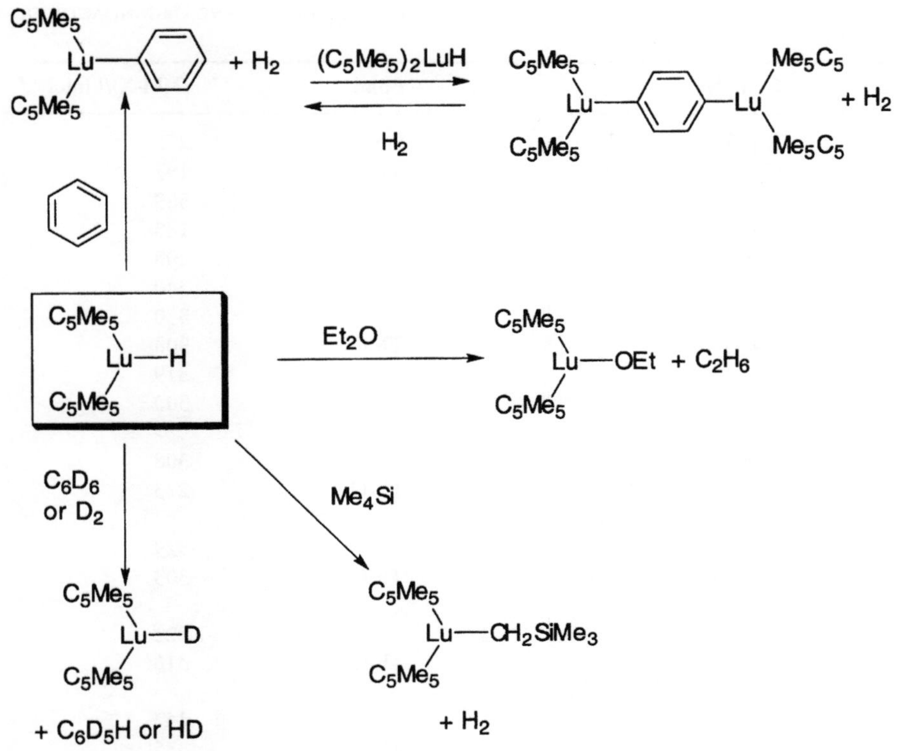

SCHEME 4.11 REACTIONS OF $[(C_5Me_5)_2LuH]$.

of these is summarized in Table 4.1. D(Th-C) for Cp_3ThMe is defined as DH for the reaction:

$$Cp_3Th - Me \rightarrow Cp_3Th + Me\bullet$$

which involves reduction of Th(IV) to Th(III). The values have been estimated from calorimetry of alcoholysis reactions such as:

$$Cp^*{}_2ThMe_2 + Bu^tOH \rightarrow Cp^*{}_2ThMe(OBu^t) + CH_4$$

In this case two separate D(M-C) values are obtained: one for $Cp^*{}_2ThMe_2$ and a second one for $Cp^*{}_2ThMe(OBu^t)$.

It is immediately apparent that all of the Ln/An-alkyl and Ln/An-hydride bonds are quite strong compared with those of middle to late transition metals. For pairs of analogous compounds D(Th-R) > D(U-R) as illustrated by comparison of entries 6, 16 and 12, 17 in Table 4.1. The homolytic cleavage which defines D(M-R) results in reduction of M; for the An(IV) complexes in Table 4.1 this would result in reduction to An(III). The increased stability of U(III) compared with Th(III) results in a less endothermic homolysis reaction for U than for Th, and thus to a smaller

TABLE 4.1 BOND DISRUPTION ENTHALPIES FOR ORGANOLANTHANIDE AND ORGANOACTINIDE COMPLEXES.

Entry	Complex	Bond	D(M-X)/kJmol^{-1},#
1	$[Cp_2Sm(\mu\text{-}H)]_2$	Sm-H	218
2	$Cp_2SmCH(SiMe_3)_2$	Sm-C	197
3	$[Cp_2SmCCPh]_2$	Sm-C	389
4	$Cp_2Sm(\eta^3\text{-}C_3H_5)$	Sm-allyl	188
5	$[Cp^*_2ThH(\mu\text{-}H)]_2$	Th-H	408
6	$Cp^*_2ThMe_2$	Th-C	340
			350
7	$Cp^*_2ThEt_2$	Th-C	308
			319
8	$Cp^*_2Th(Bu^n)_2$	Th-C	300
9	Cp*, Cp*, Th, Me, Me (structure)	Th-C	308
			273
			329
10	$Cp^*_2Th(CH_2CMe_3)_2$	Th-C	303
11	Cp*, Cp*, Th, Si, Me, Me (structure)	Th-C	322
			316
			347
12	$Cp^*_2Th(CH_2SiMe_3)_2$	Th-C	335
			344
13	Cp_3ThMe	Th-C	375
14	$Cp_3ThCH_2CMe_3$	Th-C	342
15	$Cp_3ThCH_2SiMe_3$	Th-C	368
16	$Cp^*_2UMe_2$	U-C (average)	300
17	$Cp^*_2U(CH_2SiMe_3)_2$	U-C (average)	307

Where two values are given, these refer to the cleavage of the first and second M-C bonds respectively. The values for $Cp^*_2UR_2$ are an average for the two U-C bonds.
Values relate to species in solution and are taken from Bruno, Marks and Morss (1983); Sonnenberger, Morss and Marks (1985); Bruno *et al.* (1986), and Nolan, Stern and Marks (1989).

bond disruption enthalpy for U than for Th. The larger value for the second D(M-R) compared with the first for $Cp^*_2ThR_2$ is explained by the extra stability conferred on Th(IV) by the hard OBu^t ligand of $Cp^*_2ThMe(OBu^t)$. The larger value of D(Th-R) for Cp_3ThR complexes compared with $Cp^*_2ThR_2$ (entries 6, 10, 12, 13, 14 and 15) has still not been explained; the greater steric crowding of Cp_3ThR compared with $Cp^*_2ThR_2$ might have been expected to result in the opposite trend. It is, however, apparent that D(M-R) is affected by the nature of the ancillary ligands.

The difference D(M-H) – D(M-C) is smaller by about 40 kJmol^{-1} for Ln and An complexes than for those of middle to late transition metals. The reason appears to be related to the high polarity of the Ln/An-to-ligand bond and the inability of the non-polarizable hydride ligand to stabilize negative charge. The result is that β-hydride elimination is a more endothermic process for f-block alkyls than for those of d-transition metals, and only occurs when there is steric destabilization of the alkyl or an ultimate thermodynamic sink. In alkene polymerizations the unfavourablility of the chain-terminating β-hydride elimination compared with chain propagating alkene insertion means that Ln and An catalysts have considerable thermodynamic advantages over transition metal analogues.

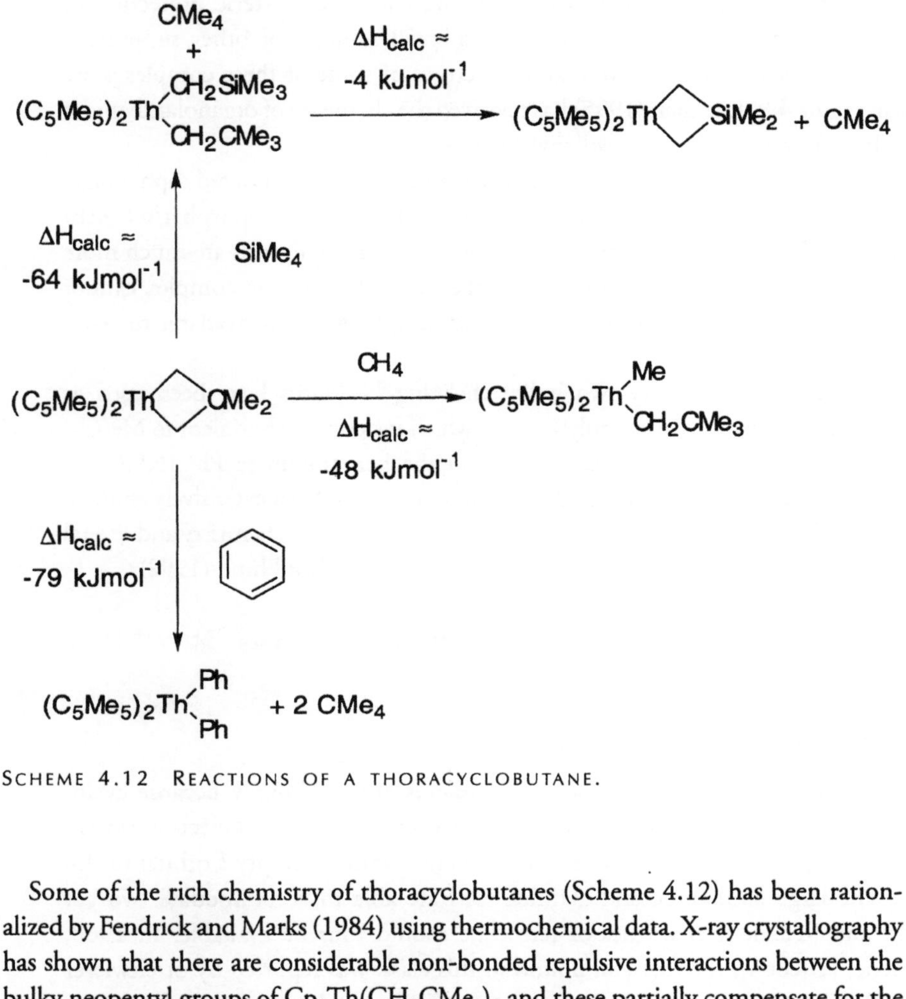

CMe_4 + $(C_5Me_5)_2Th(CH_2SiMe_3)(CH_2CMe_3)$ → ($\Delta H_{calc} \approx$ -4 $kJmol^{-1}$) → $(C_5Me_5)_2Th(CH_2)_2SiMe_2$ + CMe_4

$SiMe_4$ ($\Delta H_{calc} \approx$ -64 $kJmol^{-1}$)

$(C_5Me_5)_2Th(CH_2)_2CMe_2$ → CH_4 ($\Delta H_{calc} \approx$ -48 $kJmol^{-1}$) → $(C_5Me_5)_2Th(Me)(CH_2CMe_3)$

benzene ($\Delta H_{calc} \approx$ -79 $kJmol^{-1}$)

$(C_5Me_5)_2ThPh_2$ + 2 CMe_4

SCHEME 4.12 REACTIONS OF A THORACYCLOBUTANE.

Some of the rich chemistry of thoracyclobutanes (Scheme 4.12) has been rationalized by Fendrick and Marks (1984) using thermochemical data. X-ray crystallography has shown that there are considerable non-bonded repulsive interactions between the bulky neopentyl groups of $Cp_2Th(CH_2CMe_3)_2$ and these partially compensate for the considerable ring strain in the thoracyclobutane which is formed by elimination of CMe_4 (Scheme 4.12). The cyclometallation reaction is still calculated to be endothermic ($\Delta H_{calc} \approx 29$ $kJmol^{-1}$) but this is outweighed by entropic factors. Elimination of $SiMe_4$ from $Cp_2Th(CH_2SiMe_3)_2$ to give the corresponding Si-containing thoracyclobutane is calculated to be less endothermic, partly due to reduced ring-strain. The ring strain in these thoracyclobutanes is manifested in their anomalously low values for the first D(Th-C) (entries 9 and 11 in Table 4.1). Scheme 4.12 shows three examples of facile

C-H activation by a strained thoracyclobutane, all of which are calculated to be exothermic. None of these reactions has been observed for the analogous Si containing thoracyclobutane which is less strained, and in these cases themochemical calculations predict the reactions to be significantly less exothermic.

4.6 OTHER SUPPORTING LIGANDS FOR ORGANO-LANTHANIDE AND -ACTINIDE CHEMISTRY

Cyclopentadienyls are the most important supporting ligands for organo-lanthanide and -actinide chemistry, largely because their size allows some steric protection of the large metal centre. There are, however, a small number of other supporting ligands which have been employed, and structures of some of these complexes are shown in Figure 4.9. Edelman (1995) has reviewed the chemistry of organolanthanides which do not contain cyclopentadienyl ligands.

For lanthanides these include alkoxides such as the sterically hindered diphenolate reported by Schaverien, Meijboom and Orpen (1992) and octaethylporphyrin (OEP) reported by Schaverien and Orpen (1991). These supporting ligands are much more limited in their use than cyclopentadienyls: the bulky diphenolate complex cannot be isolated for smaller lanthanides, and the OEP complexes are not available for early lanthanides.

In organoactinide chemistry two bulky chelating N-donors have been used as supporting ligands: hydrotris(pyrazolyl)borate, which is sterically equivalent to $Me_5C_5^-$ has been used by Domingos *et al.* (1994) and the benzamidinate $PhC(NSiMe_3)_2$ reported by Wedler *et al.* (1992), but their chemistry has not been extensively studied. The bulky $N(SiMe_3)_2$ ligand has also been used in organoactinide chemistry and several reactions of $[UMe\{N(SiMe_3)_2\}_3]$ have been reported by Ephritikhine (1992).

- Organolanthanide and -actinide chemistry with supporting ligands other than Cp is very limited.

4.7 CYCLO-OCTATETRAENYL COMPLEXES

Interest in f-element cyclo-octatetraenyl complexes arose primarily because cyclo-octatetraenyl, $C_8H_8^{2-}$ (abbreviated as cot) is a planar, aromatic 10 π electron system with a HOMO of the correct symmetry to interact with an empty f orbital on Ln or An. The large lanthanide and actinide ions are able to accommodate two cot ligands, and f-element analogues of ferrocene should thus be available. In 1968, uranocene, $[U(cot)_2]$, the first compound in this class, was reported by Streitwieser and Muller-Westhof. The related Ln complexes $K[Ln(cot)_2]$ were reported one year later. Since the first reports, $[An(cot)_2]$ have also been prepared for An = Th, Np and Pu, and a range of complexes with substituted analogues of cot have been made.

The synthesis of the complexes is by the metathesis reaction of the appropriate halide of Ln or An with K_2cot as shown below.

$$UCl_4 + 2K_2C_8H_8 \rightarrow [U(C_8H_8)_2] + 4KCl$$

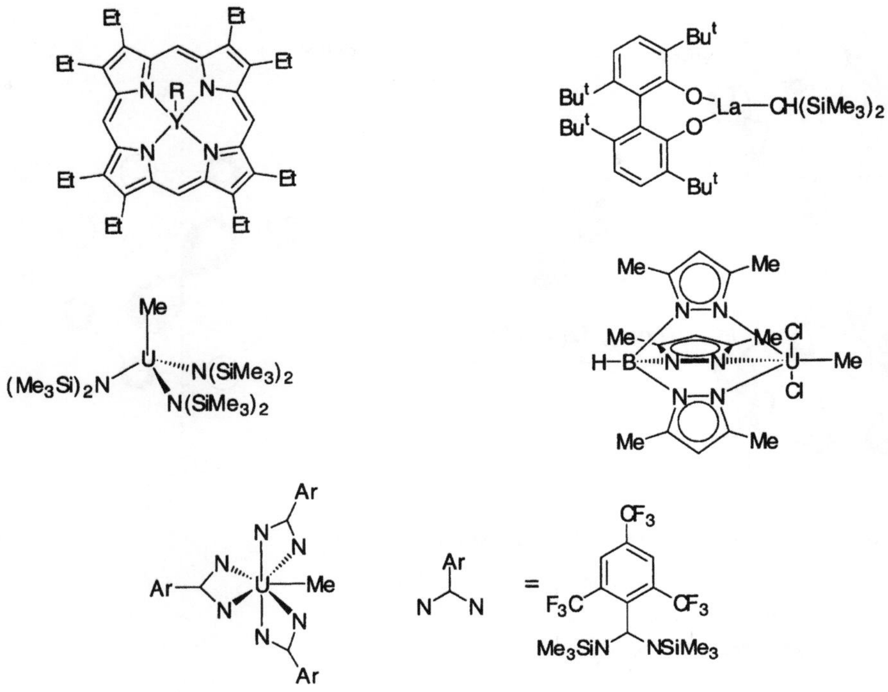

FIGURE 4.9 ORGANOLANTHANIDE AND ACTINIDE COMPLEXES WITH NON-CP SUPPORTING LIGANDS.

Not surprisingly, a large proportion of the publications on these complexes relate to structure and bonding. Uranocene was shown by X-ray diffraction to have a sandwich structure with planar, eclipsed cot rings and overall D_{8h} symmetry, and a qualitative bonding picture was presented soon afterwards by Streitwieser *et al.* (1973). In this picture the e_{2u} HOMO's on cot interact with the empty f(xyz) and f(z(x^2–y^2)) orbitals on U which transform as e_{2u} in the D_{8h} symmetry of the eclipsed molecule as shown in Figure 4.10.

It should be noted, however that the crystal structure of [U(η^8-C_8Me_8)$_2$] shows two rotamers, one with eclipsed and one with staggered rings. Calculations suggest that the f-orbital contribution to bonding in $[Ce(cot)_2]^-$ is only one seventh that in uranocene, and the cot rings in $[Ln(cot)_2]^-$ complexes are found to be staggered rather than eclipsed. The crystal structures of [K(diglyme)][Ce(cot)$_2$] and uranocene are shown in Figure 4.11.

Uranocene is a thermally stable compound which can be sublimed without decomposition above 200°C. Although it reacts rapidly with oxygen and is pyrophoric, its chemistry is consistent with significant covalent contribution to bonding: it is not susceptible to ligand exchange and it reacts only slowly with protic reagents such as H_2O and acetic acid. In contrast $[Ln(cot)_2]^-$ are hydrolyzed instantly in the presence of even small quantities of H_2O. Another illustration of the greater stability

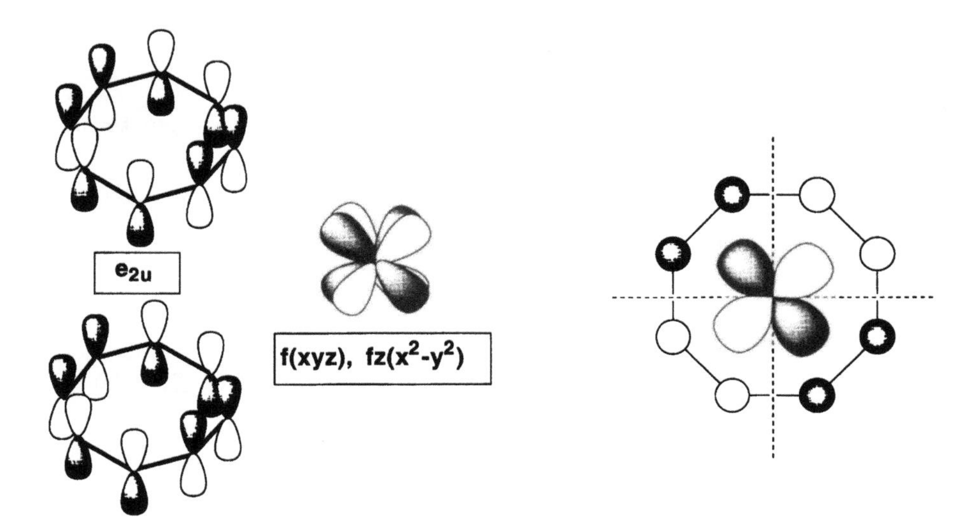

FIGURE 4.10 MOLECULAR ORBITALS FOR URANOCENE.

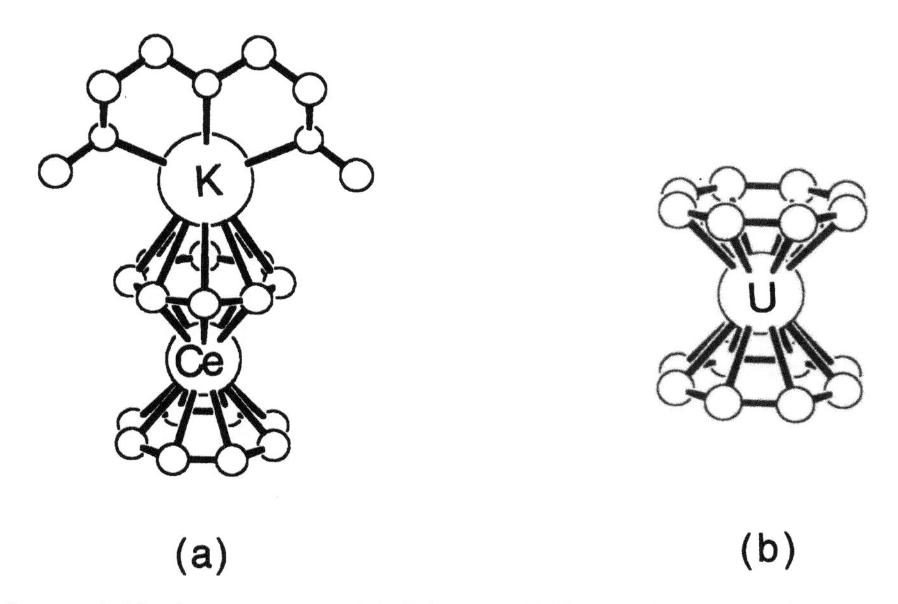

FIGURE 4.11 STRUCTURES OF (A) $[K(DIGLYME)][Ce(COT)_2]$ AND (B) $[U(COT)_2]$.

of uranocene compared with $[Ln(cot)_2]^-$ is the reaction between $K[Ln(cot)_2]$ and UCl_4 which proceeds rapidly and quantitatively to give KCl, $LnCl_3$ and $[U(cot)_2]$.

Careful reaction of $[Ce(OPr^i)_4{\cdot}Pr^iOH]$ with $AlEt_3$ in C_8H_8 results in formation of the highly air-sensitive brown-black compound cerocene, $[Ce(cot)_2]$. Its formula suggests that Ce is in oxidation state +4, however the strongly oxidizing Ce(IV) would not be expected to coexist with the reducing cot ligand, and the alternative formulation $[Ce^{3+}\{(cot)_2\}^{3-}]$ has been suggested. This formulation is supported by *ab initio* calculations which show that the e_{2u} ligand HOMO interacts with the e_{2u} Ce 4f orbital to give a singlet ground state $4f(e_{2u})^1\pi(e_{2u})^3$. This cannot be verified by magnetic measurements, but has been confirmed by Edelstein *et al.* (1996) using X-ray absorption near edge structure (XANES). Cerocene reacts relatively slowly with H_2O, but calculations indicate that covalent contributions to the bonding are less than in uranocene. Early work on lanthanide and actinide cot complexes has been reviewed by Marks and Ernst (1982).

- $[U(cot)_2]$ has D_{8h} symmetry with eclipsed cot rings.
- $[Ln(cot)_2]^-$ have staggered cot rings.
- HOMO's of cot have the correct symmetry (e_{2u}) in a D_{8h} complex to interact with metal f(xyz) and $f(z(x^2-y^2))$ orbitals.
- $[U(cot)_2]$ is much less reactive than $[Ln(cot)_2]^-$, consistent with some covalent contribution to the bonding in $[U(cot)_2]$.

4.8 ARENE COMPLEXES

Arene complexes are well known in organotransition metal chemistry, but were much later on the scene in organo-f element chemistry. F-element halides were known to form intercalation compounds with graphite and so it was reasonable to expect that complexes of these halides with benzene should be available. The first example to be structurally characterized was $[U(\eta^2\text{-}AlCl_4)_3(\eta^6\text{-}C_6H_6)]$, prepared by Cesari *et al.* (1971) by reaction of UCl_4, $AlCl_3$ and Al shavings in refluxing C_6H_6, and an analogous complex of Sm(III) was reported some years later by Cotton and Schwotzer (1986). The structure of $[U(\eta^2\text{-}AlCl_4)_3(\eta^6\text{-}C_6H_6)]$ is shown in Figure 4.12. η^6 arene interactions have been observed in the aryloxide complex $[Nd(O\text{-}2,6\text{-}Pr^i_2C_6H_3)_2(\eta^6\text{-}\mu_2\text{-}O\text{-}2,6\text{-}Pr^i_2C_6H_3)]_2$ (Chapter 3) and in diphenylphenoxide complexes of Eu and Yb reported by Deacon *et al.* (1999).

Zero-valent Lanthanide Arene Complexes

$[Ln(\eta^6\text{-}1,3,5\text{-}Bu^t_3\text{-}C_6H_3)_2]$ reported by Brennan *et al.* (1987) were the first examples of lanthanide complexes in oxidation state zero. They were prepared by reaction of metal vapour with the bulky arene 1,3,5-tri-t-butylbenzene, and were isolated and characterized for all the lanthanides except Ce, Eu, Tm and Yb. The structure of the Gd complex is shown in Figure 4.13.

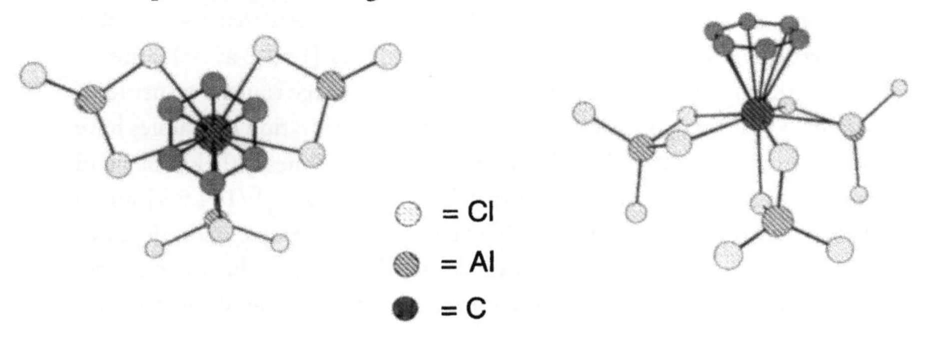

FIGURE 4.12 TWO VIEWS OF THE STRUCTURE OF $[U(\eta^2\text{-}AlCl_4)_3(\eta^6\text{-}C_6H_6)]$.

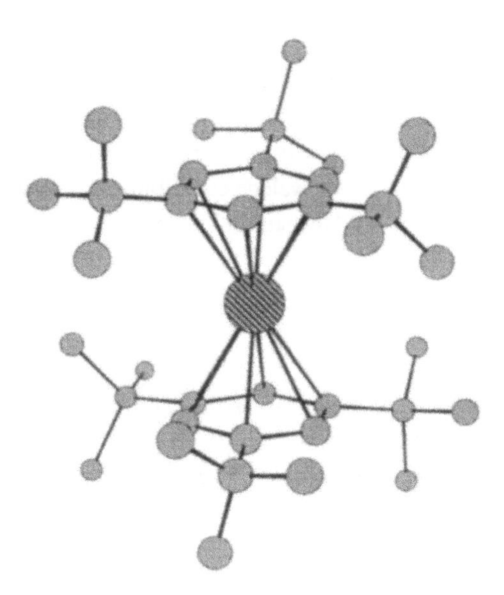

FIGURE 4.13 STRUCTURE OF $[Gd(\eta^6-C_6H_3Bu^t_3)_2]$.

The bonding in these complexes has been analyzed by analogy with that in the 18-electron $[Cr(\eta^6-C_6H_6)_2]$, and can be understood in terms of delocalization of metal-based e_{2g} electrons onto the arene rings as shown in Figure 4.14. The 15-electron complex $Y(\eta^6-1,3,5-Bu^t_3-C_6H_3)_2]$ has three electrons in the degenerate e_{2g} orbitals and thus has a ^2E ground state which has been characterized by low temperature esr spectroscopy. A stable $[Ln(\eta^6\text{-arene})_2]$ requires an easily accessible d^1s^2 state and the stability of these complexes along the lanthanide series has been correlated with the f^ns^2 to $f^{n-1}d^1s^2$ transition energies. The very high transition energies for Ln = Sm, Eu, Tm, Yb due to the stability of $f^{6,7,13,14}$ configurations result in instability of bis(arene) complexes for these elements. The La complex decomposes above 0°C; its instability and the impossibility of isolating a complex for Ce have been ascribed to the large size of the metal atoms. Theoretical studies suggest that the 4f-orbitals make essentially no contribution to the bonding, however the observation that the magnetic moments of $[Ln(\eta^6\text{-arene})_2]$ containing unpaired f electrons are very different from those of the free atoms suggests that the f-shell is significantly perturbed. Unlike most lanthanide complexes $[Ln(\eta^6\text{-arene})_2]$ are all highly coloured; this has been ascribed to ligand to metal charge transfer transitions. The Ln to arene bond is by no means weak: mean bond dissociation enthalpies have been measured for $[Ln(\eta^6-1,3,5-Bu^t_3-C_6H_3)_2]$. For Ln = Y, the value is 301 $kJmol^{-1}$ and for Dy it is 197 $kJmol^{-1}$, compared with those for $[Cr(\eta^6-C_6H_6)_2]$ (164.9 $kJmol^{-1}$) and $[W(\eta^6-C_6H_5Me)_2]$ (303.9 $kJmol^{-1}$). No zero-valent bis(arene) actinide complexes have yet been prepared, but theoretical studies by Hong, Schautz and Dolg (1999) suggest that they should be stable compounds and that the 5f orbitals may contribute to the bonding.

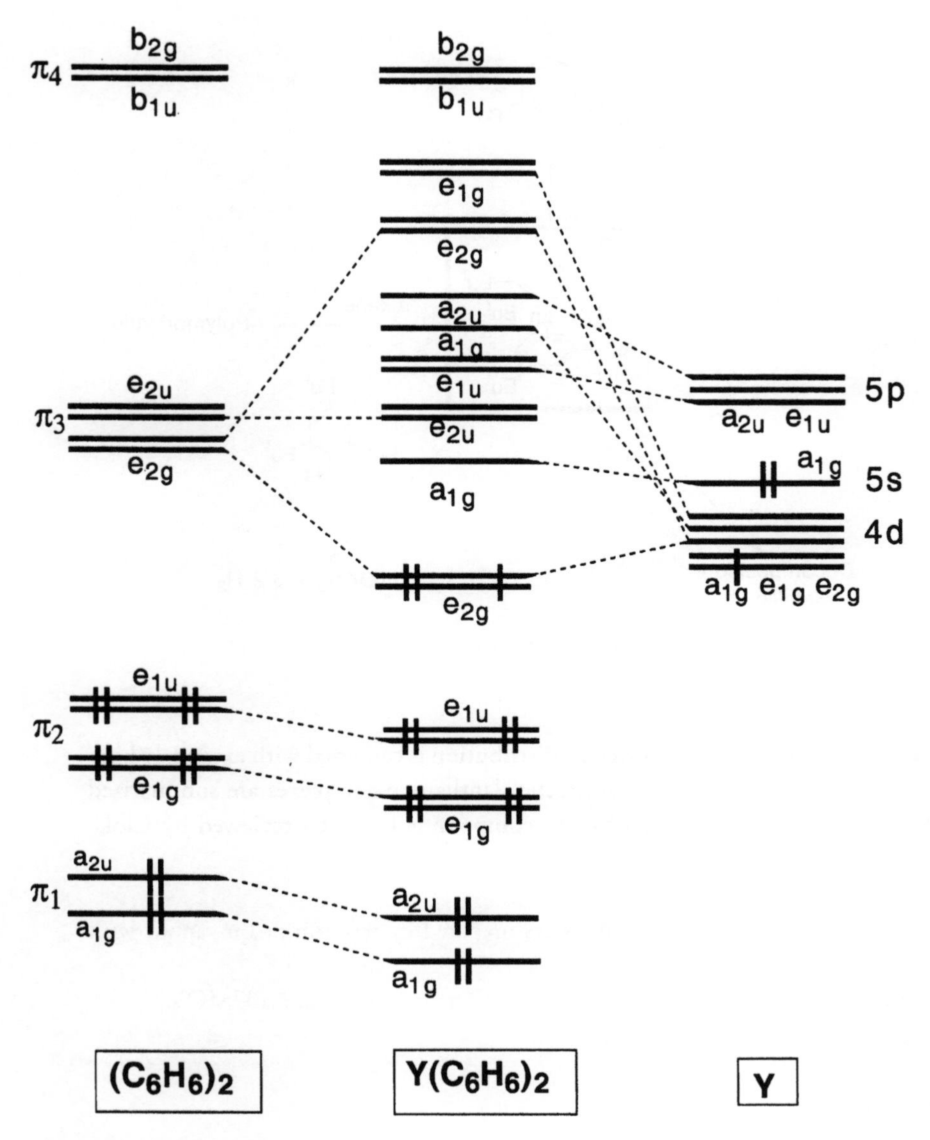

FIGURE 4.14 MOLECULAR ORBITAL SCHEME FOR $[Y(\eta^6\text{-}C_6H_6)_2]$.

As expected, all the lanthanide bis(arene) complexes are extremely air and moisture sensitive and they readily undergo redox reactions with protic reagents such as phenols and secondary amines to produce Ln(III) complexes. Reaction with CO or PR_3 results only in decomposition and precipitation of metallic Ln; only one straightforward ring displacement reaction, with the strong π acceptor $Bu^tNCHCHNBu^t$, has been reported, and the product of this reaction may be best described as $Ln^{3+}[(Bu^tNCHCHNBu^t)^{\cdot-}]_3$ in which the ligands are radical anions. Probably the most exciting reactivity is that with alkenes where polymerization to give high density

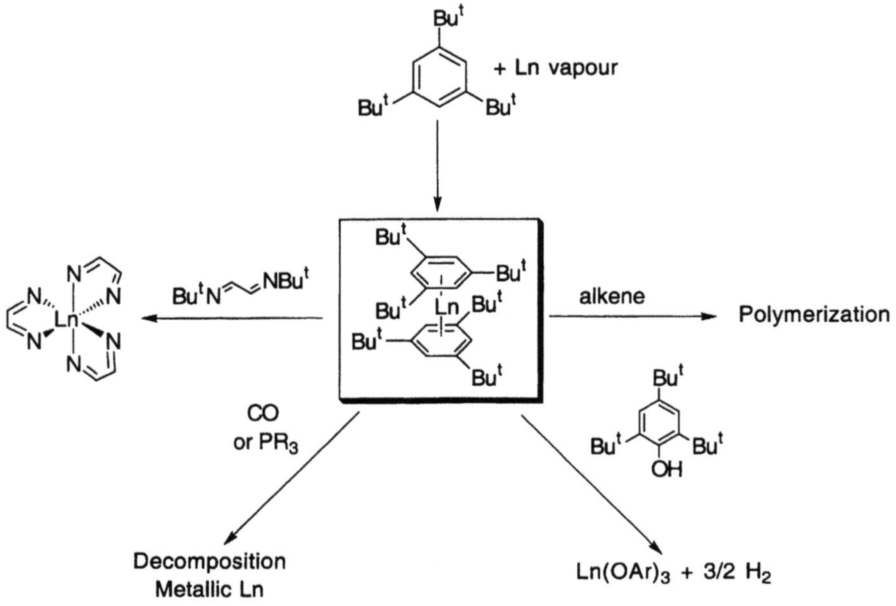

SCHEME 4.13 REACTIONS OF $[Ln(\eta^6\text{-}C_6H_3Bu^t{}_2)_2]$

polymers with narrow molecular weight distribution is catalyzed with extremely high turnovers. Important reactions of bis(arene) lanthanide complexes are summarized in Scheme 4.12. The chemistry of these compounds has been reviewed by Cloke (1993).

- Zero-valent Ln bis(arene) complexes are prepared by reaction of Ln vapour with arene.
- Bonding is through donation of Ln d^1s^2 electrons into arene LUMO's.
- Most reactions involve oxidation of Ln.

REFERENCES

Boncella, J.M., Tilley, T.D. and Andersen, R.A. (1984) Reaction of $M(C_5Me_5)_2(OEt_2)$ M = Eu or Yb with phenylacetylene; formation of mixed valence $Yb_3(C_5Me_5)_4(m\text{–}C{\equiv}CPh)_4$ and $Eu_2(C_5Me_5)_2(m\text{–}C{\equiv}CPh)_2(tetrahydrofuran)_4$. *J. Chem. Soc., Chem. Commun.*, 710–712.

Brennan, J.G., Andersen, R.A. and Robbins, J.L. (1986) Preparation of the first molecular carbon–monoxide complex of uranium, $(Me_3SiC_5H_4)_3UCO$. *J. Am. Chem. Soc.*, **108**, 335–336.

Brennan, J.G., Cloke, F.G.N., Sameh, A.A. and Zalkin, A. (1987) Synthesis of Bis(η^6-1,3,5-tri-t-butylbenzene) Sandwich Complexes Of Yttrium(0) and Gadolinium(0) – The X-Ray Crystal- Structure of the 1st Authentic Lanthanide(0) Complex, $[Gd(\eta\text{-}Bu^t{}_3C_6H_3)_2]$ *J. Chem. Soc., Chem. Commun.*, 1668–1669.

Bruno, J.W., Marks, T.J. and Morss, L.R. (1983) Organo-f-element thermochemistry. Metal-ligand bond disruption enthalpies in (pentamethylcyclopentadienyl)thorium hydrocarbyls, metallacycles, hydrides, and dialkylamides. *J. Am. Chem. Soc.*, **105**, 6824–6832.

Bruno, J.W., Smith, G.M., Marks, T.J., Fair, C.K., Schulz, A.J. and Williams, J.M. (1986) C–H activation mechanisms and regioselectivity in the cyclometalation reactions of bis(pentamethylcyclopentadienyl) thorium dialkyl complexes. *J. Am. Chem. Soc.*, **108**, 40–56.

Bruno, J.W., Stecher, H.A., Morss, L.R., Sonnenberger, D.C. and Marks, T.J. (1986) Organo-f-element thermochemistry. Thorium vs. uranium and ancillary ligand effects on metal-ligand bond disruption enthalpies in bis(pentamethylcyclopentadienyl)actinide bis(hydrocarbyls) and bis(pentamethylcyclopentadienyl)alkoxyactinide hydrides and hydrocarbyls. *J. Am. Chem. Soc.*, **108**, 7275–7280.

Bursten, B.E. and Strittmatter, R.J. (1991) Cyclopentadienyl-actinide complexes: bonding and electronic structure. *Angew. Chem. Int. Ed.*, **30**, 1069–1085.

Cesari, M., Pedretti, U., Zazzetta, A., Lugli, G. and Marconi, W. (1971) Synthesis and Structure of a π-Arene Complex of Uranium(III)–Aluminium Chloride. *Inorg. Chim Acta*, **5**, 439–444.

Cloke, F.G.N. (1993) Zero Oxidation State Compounds of Scandium, Yttrium and the Lanthanides. *Chem. Soc. Reviews.*, **22**, 17–24.

Conejo, M.D., Parry, J.S., Carmona, E., Schultz, M., Brennan, J.G., Beshouri, S.M., Andersen, R.A., Rogers, R.D., Coles, S. and Hursthouse, M. (1999) Carbon monoxide and isocyanide complexes of trivalent uranium metallocenes. *Chemistry – A European Journal*, **5**, 3000–3009.

Cotton, F.A. and Schwotzer, W. (1986) $Sm(\eta^6–C_6Me_6)(\eta^2–AlCl_4)_3$: The First Structure of a Rare Earth Complex with a Neutral π-Ligand. *J. Am. Chem. Soc.*, **108**, 4657–4658.

Deacon, G.B., Forsyth, C.M., Junk, P.C., Skelton, B.W. and White, A.H. (1999) The striking influence of intramolecular lanthanide — π–arene interactions on the structure of the homoleptic aryloxolanthanoid(II) complexes $\{Eu_2(Odpp)(\mu\text{-}Odpp)_3]$ and $[Yb2(Odpp)(\mu\text{-}Odpp)_3]$ and the Yb(II)/Yb(III) trimetallic $[Yb_2(\mu\text{-}Odpp)_3]^+[Yb(Odpp)4]^-(^-Odpp$ = 2,6-diphenylphenolate). *Chem. Eur. J.*, 1452–1459.

Di Bella, S., Lanza, G., Fragalà, I.L. and Marks, T.J. (1996) Electronic structure and photoelectron spectroscopy of the monomeric uranium(III) alkyl $[\eta^5–(CH_3)_5C_5]_2UCH[(Si(CH_3)_3]_2$. *Organometallics*, **15**, 205–208.

DiBella, S., Gulino, A., Lanza, G., Fragala, I., Stern, D. and Marks, T.J. (1994) Photoelectron spectroscopy of f–element organometallic complexes. 12. A comparative investigation of the electronic structure of lanthanide bis(polymethylcyclopentadienyl)hydrocarbyl complexes by relativistic *ab initio* and DV–Xα calculations and gas–phase UV photoelectron spectroscopy. *Organometallics*, **13**, 3810–3815.

Domingos, A., Marquez, N., Dematos, A.P., Santos, I. and Silva, M. (1994) Hydrotris(pyrazolyl)borate Chemistry of Uranium(III) and Uranium(IV) — Synthesis of σ-Hydrocarbyl Derivatives of U(IV) and Reactivity of $UCl_2R[HB(3,5\text{-}Me_2Pz)_3]$ (R = CH_2SiMe_3, $CH(SiMe_3)_2$) and $UCl_2[HB(3,5\text{-}Me_2Pz)_3]$ Towards Ketones and Aldehydes.*Organometallics*, **13**, 654–662.

Eaborn, C., Hitchcock, P.B., Izod, K. and Smith, J.D. (1994) A monomeric solvent free bent lanthanide dialkyl and a lanthanide analogue of a Grignard reagent. Crystal structures of $Yb\{C(SiMe_3\}_2$ and $[Yb\{C(SiMe_3)_3\}IOEt_2]_2$. *J. Amer. Chem Soc.*, **116**, 12071–12072.

Eaborn, C., Hitchcock, P.B., Izod, K. and Smith, J.D. (1996) Alkyl derivatives of europium (+2) and ytterbium (+2). Crystal structures of $Eu(C(SiMe_3)_3]_2$, $Yb[C(SiMe_3)_2(SiMe_2\ CH{=}CH_2)]IOEt_2$ and $Yb[C(SiMe_3)_2(SiMe_2OMe)]IOEt_2$. *Organometallics*, **15**, 4783–4790.

Edelmann, F.T. (1995) Cyclopentadienyl-Free Organolanthanide Chemistry. *Angew. Chem., Int. Ed.*, **34**, 2466–2488.

Edelstein, N.M., Allen, P.G., Bucher, J.J., Shuh, D.K., Sofield, C.D., Kaltsoyannis, N., Maunder, G.H., Russo, M.R. and Sella, A. (1996) The Oxidation State of Ce in the Sandwich Molecule Cerocene *J. Am. Chem. Soc.*, **118**, 13115–13116.

Ephritikhine, M. (1992) Recent Advances in Organoactinide Chemistry. *New Journal of Chemistry*, **16**, 451–469.

Evans, W.J., Engerer, S.C. and Coleson, K.M. (1981) Reactivity of lanthanide metals with unsaturated–hydrocarbons – terminal alkyne reactions. *J. Am. Chem. Soc.*, **103**, 6672–6777.

Evans, W.J., Meadows, J.H., Wayda, A.L., Hunter, W.E. and Atwood, J.L. (1982) Organolanthanide hydride chemistry. 1. Synthesis and X-ray crystallographic characterization of dimeric organolanthanide and organoyttrium hydride complexes. *J. Am. Chem. Soc.*, **104**, 2008–2014.

Evans, W.J., Ulibarri, T.A. and Ziller, J.W. (1988) Isolation and X–ray crystal structure of the first dinitrogen complex of an f–element metal, $[(C_5Me_5)_2Sm]_2N_2$. *J. Am. Chem. Soc.*, **110**, 6877–6879.

Fendrick, C.M. and Marks, T.J. (1984) Thermochemically based strategies for C–H activation on saturated hydrocarbon molecules. Ring opening reactions of a thoracyclobutane with tetramethylsilane and methane. *J. Am. Chem. Soc.*, **106**, 2214–2216.

Giardello, M.A., Conticello, V.P., Brard, L., Sabat, M., Rheingold, A.L., Stern, C.L. and Marks, T.J. (1994) Chiral Organolanthanides Designed for Asymmetric Catalysis-Synthesis, Characterization, and Configurational Interconversions Of Chiral, C-1-Symmetrical Organolanthanide Halides, Amides, and Hydrocarbyls. *J. Am. Chem. Soc.*, **116**, 10212–10240.

Haar, C.M., Stern, C.L. and Marks, T.J. (1996) Coordinative unsaturation in chiral organolanthanides. Synthetic and asymmetric catalytic mechanistic study of organoyttrium and –lutetium complexes having pseudo–meso $Me_2Si(\eta^5\text{-}RC_5H_3)(\eta^5\text{-}R(^*)C_5H_3)$ ancillary ligation. *Organometallics*, **15**,.1765–1784.

Hitchcock, P.B., Lappert, M.F., Smith, R.G., Bartlett, R.A. and Power, P.P. (1988) Synthesis and structural characterization of the 1st neutral homoleptic lanthanide metal(III) alkyls–$[LnR_3]$ [Ln = La or Sm, R = $CH(SiMe_3)_2$]. *J. Chem. Soc. Chem. Commun.*, 1007–1009.

Hong, G., Schautz, F. and Dolg, M. (1999) *Ab Initio* Study of Metal-Ring Bonding in the Bis(η^6–benzene)lanthanide and -actinide Complexes $M(C_6H_6)_2$ (M = La, Ce, Nd, Gd, Tb, Lu, Th, U). *J. Am. Chem. Soc.*, **121**, 1502–1512.

Marks, T.J. and Ernst, R.D. (1982) Scandium, Yttrium, the Lanthandes and Actinides. In *Comprehensive Organometallic Chemistry* edited by Wilkinson, G., Stone, F.G.A., Abel, E.W., Vol. 3. Pergamon Press.

Marks, T.J. and Grynkewich, G.W. (1976) Organolanthanide tetrahydroborates. Ligation geometry and coordinative saturation. *Inorg. Chem.*, **15**, 1302–1314.

Nolan, S.P., Stern, D. and Marks, T.J. (1989) Organo-f-element thermochemistry. Absolute metal–ligand bond disruption enthalpies in bis(pentamethylcyclopentadienyl)samarium hydrocarbyl, hydride, dialkylamide, alkoxide, halide, thiolate and phosphide complexes. Implications for organolanthanide bonding and reactivity. *J. Am. Chem. Soc.*, **111**, 7844–7853.

Schaverien, C.J., Meijboom, N. and Orpen, A.G. (1992) A New Ligand Environment in Organolanthanoid Chemistry: Sterically Hindered, Chelating Diolato Ligands and the X-ray Structure of $[La\{CH(SiMe_3)_2\}\{1,1'\text{-}(2\text{-}OC_6H_2Bu^t{}_2\text{-}3,5)_2(THF)_3]$. *J. Chem. Soc., Chem. Commun.*, 124–126.

Schaverien, C.J. and Orpen, A.G. (1991) Chemistry of (Octaethylporphyrinato)lutetium and -yttrium Complexes: Synthesis and Reactivity of (OEP)MX Derivatives and the Selective Activation of O_2 by (OEP)Y(μ-Me)$_2$AlMe$_2$. *Inorg. Chem.*, **30**, 4968–4978.

Schock, L.E., Seyam, A.M., Sabat, M. and Marks, T.J. (1988) A new approach to measuring absolute metal–ligand bond disruption enthalpies in organometallic compounds. The $[(CH_3)_3SiC_5H_4]_3U$ system. *Polyhedron*, 7, 1517–1529.

Schumann, H. and Genthe, W. (1981) Organometallic compounds of lanthanoids .77. Dicyclopentadienyllutetium hydride. *J. Organomet. Chem.*, **213**, C7–C9.

Schumann, H., Meese–Marktscheffel, J.A. and Esser, L. (1995) Synthesis, structure, and reactivity of organometallic π-complexes of the rare earths in oxidation state Ln^{3+} with aromatic ligands. *Chem. Rev.*, **95**, 865–986.

Schumann, H., Müller, J., Bruncks, N., Lauke, H. and Pickardt, J. (1984) Organometallic Compounds of the Lanthanides. 17. Tris[(tetramethylethylenediamine)lithium] Hexamethyl Derivatives of the Rare Earths. *Organometallics*, **3**, 69–74.

Sonnenberger, D.C., Morss, L.R. and Marks, T.J. (1985) Organo-f-element thermochemistry. Thorium-ligand bond disruption enthalpies in tris(cyclopentadienyl)thorium hydrocarbyls. *Organometallics*, **4**, 352–355.

Streitwieser, A., Muller–Westerhoff, U., Sonnichsen, G., Mares, F., Morrell, D.G., Hodgson. K.O. and Harmon, C.A. (1973) Preparation and Properties of Uranocene, Di-π-cyclooctatetraeneuranium(IV). *J. Am. Chem. Soc.*, **95**, 8644–8649.

Timofeeva, T.V., Lii, J.H. and Allinger, N.L. (1995) Molecular Mechanics Explanation of the Metallocene Bent Sandwich Structure. *J. Am. Chem. Soc.*, **117**, 7452–7459.

Van der Sluys, W.G., Burns, C.J. and Sattelberger, A.P. (1989) First example of a neutral homoleptic uranium alkyl. Synthesis, properties and structure of $U[CH(SiMe_3)_2]_3$. *Organometallics*, **8**, 855–857.

Von Ammon, R., Kanellakopulos, B. and Fischer, R.D. (1970) NMR evidence for electron spin delocalization in organometallic U(IV) compounds. *Chem. Phys. Lett.*, **4**, 553–557.

Watson, P.L. (1983) Facile C-H activation by lutetium-methyl and lutetium-hydride complexes. *J. Chem. Soc. Chem. Commun.*, 276–277.

Wayda, A.L., Atwood, J.L. and Hunter, W.E. (1984) Homoleptic organolanthanoid hydrocarbyls. The synthesis and X-ray crystal structure of tris[*o*-((dimethylamino)methyl)phenyl] lutetium. *Organometallics*, **3**, 939–941.

Wedler, M., Knösel, F., Edelman, F.T. and Behrens, U. (1992) Stabilization of Uranium(IV) Alkyls by Bulky Chelating Ligands: Molecular Structure of $[PhC(N(SiMe_3)_2]_3UMe$.*Chem. Ber.*, **125**, 1313–1318.

Wilkinson, G. and Birmingham, J.M. (1954) Cyclopentadienyl compounds of Sc, Y, La, Ce and some lanthanide elements. *J. Am. Chem. Soc.*, **76**, 6210.

CHAPTER 5

LANTHANIDE COMPLEXES AS CATALYSTS AND REAGENTS FOR ORGANIC REACTIONS

Lanthanide complexes now have an established place as catalysts and reagents in organic synthesis. The gradual decrease in ionic radius of the Ln^{3+} ions on traversing the series from La to Lu allows fine-tuning of the steric properties of the metal centre and thus of the reactivity and selectivity of the complex, a possibility not available elsewhere in the periodic table. An attractive feature of lanthanides in catalysis is the high degree of lability of Ln-to-ligand bonds which allows rapid turnover rates in catalytic reactions. Many of the reactions described in this chapter can be catalyzed by non-lanthanide complexes, but usually turnover rates are much lower and sometimes (*e.g.* in Lewis acid catalysis) stoichiometric quantities of catalyst may be required. Lanthanides are not noted for the richness of their redox chemistry, but Ce(IV) compounds are widely used as oxidizing agents and SmI_2 is now an established one-electron reducing agent. The reactions described in this chapter will illustrate important aspects of lanthanide chemistry which have been encountered in earlier chapters. Although lanthanides are important components of many heterogeneous catalysts (*e.g.* CeO_2 in catalytic converters and Ln stabilized zeolites for petroleum cracking) these systems will not be discussed here. More detailed information may be found in books by Imamoto (1995) and Kobayashi (1998). The use of lanthanide catalysts in polymerization reactions has been reviewed by Yasuda and Ihara (1977).

5.1 CATALYSIS BY BIS(CYCLOPENTADIENYL) COMPLEXES

Lanthanide bis(cyclopentadienyl) complexes are the most studied organolanthanides, and their synthesis and structures have been described in Chapter 4. The possibility of fine-tuning their catalytic activity and selectivity by varying Ln and by introducing substituents into the Cp rings has resulted in a very rich catalytic chemistry for these compounds. The most active catalysts have a readily available binding site at Ln. This can be achieved by using large, early Ln, or by using sterically undemanding ligands, and reactions must be conducted in non-coordinating solvents. There is, however, a very delicate balance to be achieved, as complexes of large Ln with small ligands may well dimerize to form less active species. 'Tied-back' bis(cyclopentadienyls) where two cyclopentadienyl rings are linked by, for example, a $SiMe_2$ unit, are often

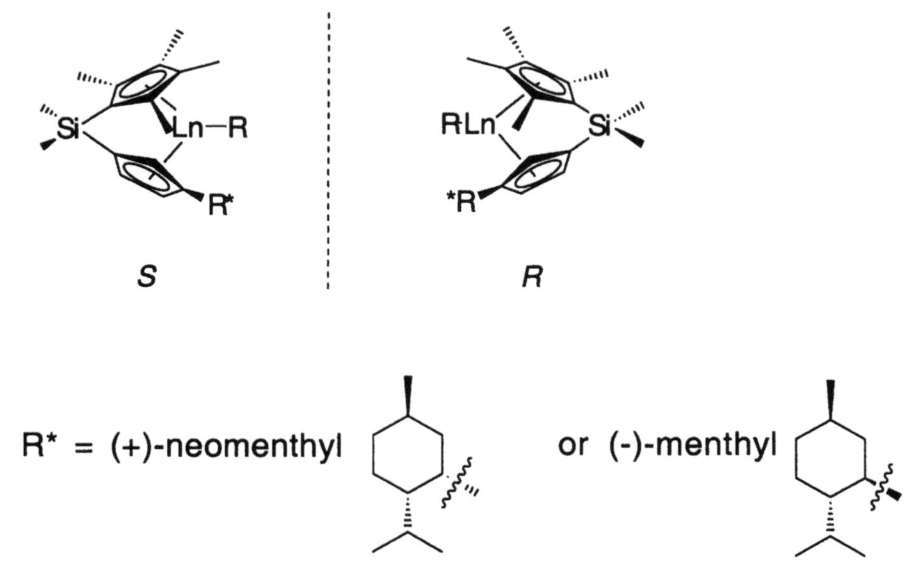

FIGURE 5.1 *S*- AND *R*- DIASTEREOMERS OF CHIRAL BIS(CYCLOPENTADIENYL) LANTHANIDE COMPLEXES.

found to show enhanced catalytic activity because the reactive Ln centre becomes more accessible to incoming substrate molecules.

Enantioselectivity is an important goal in catalytic chemistry, and as well as showing enhanced activity, these 'tied-back' bis(cyclopentadienyls) can be modified by incorporation of a chiral substituent on one of the rings to form chiral, enantioselective catalysts as reported by Giardello *et al.* (1994a). Depending on which face of the chirally substituted cyclopentadienyl ring binds to Ln, one of two diastereomers may be formed. These are designated *R* and *S* as shown in Figure 5.1.

In some reactions it is found that one diastereomer is more active and more selective than the other, *e.g.* *R*-(+)-neomenthyl may be more active than *S*-(+)-neomenthyl. Separation of the diastereomers can sometimes be achieved by fractional crystallization, but due to the differences in activity/selectivity some enantioselectivity may often be observed in reactions using diastereomerically impure catalysts. The structures of *R*-(+)-neomenthyl and *S*-(–)-menthyl samarium complexes are shown in Figure 5.2.

5.1.1 Hydrogenation of Alkenes

Lanthanide bis(pentamethylcyclopentadienyl) hydrides, usually generated by hydrogenolysis of the corresponding alkyl complex, show remarkable activity in the hydrogenation of terminal and internal alkenes. Catalytic activity increases with decreasing Ln radius and for $[(C_5Me_5)_2Lu(\mu\text{-}H)]_2$ turnovers of 120,000 h^{-1} have been observed at 25°C and 1 atmosphere of H_2, compared with turnovers of 3000 h^{-1} achieved with $[RhCl(PPh_3)_3]$ under similar conditions. The proposed mechanism for this hydrogenation reaction is shown in Scheme 5.1.

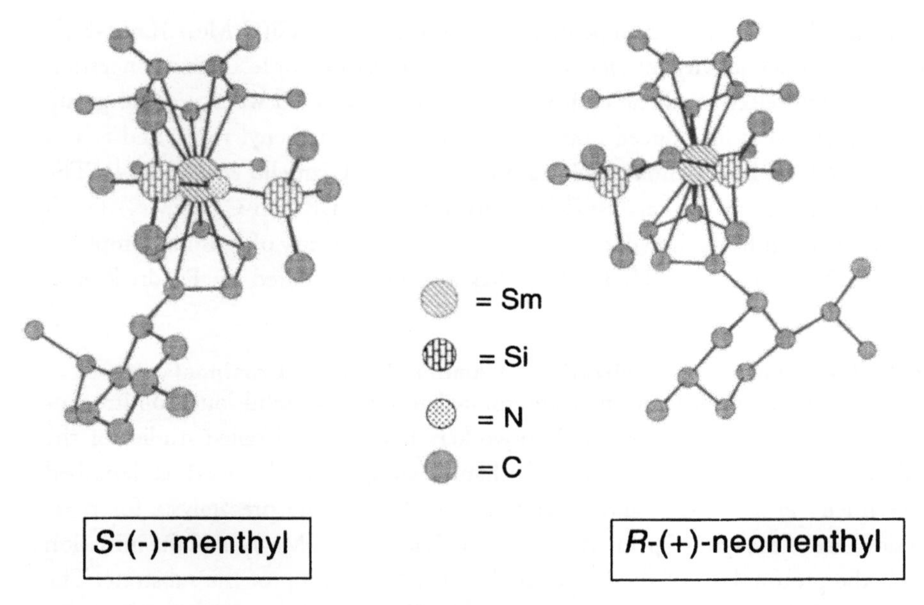

FIGURE 5.2 CRYSTAL STRUCTURES OF S-[$(C_5Me_4SiMe_2C_5H_3R^*)SmN(SiMe_3)_2$] ($R^*$ = MENTHYL) AND R-[$(C_5Me_4SiMe_2C_5H_3R^*)SmCH(SiMe_3)_2$] ($R^*$ = NEOMENTHYL).

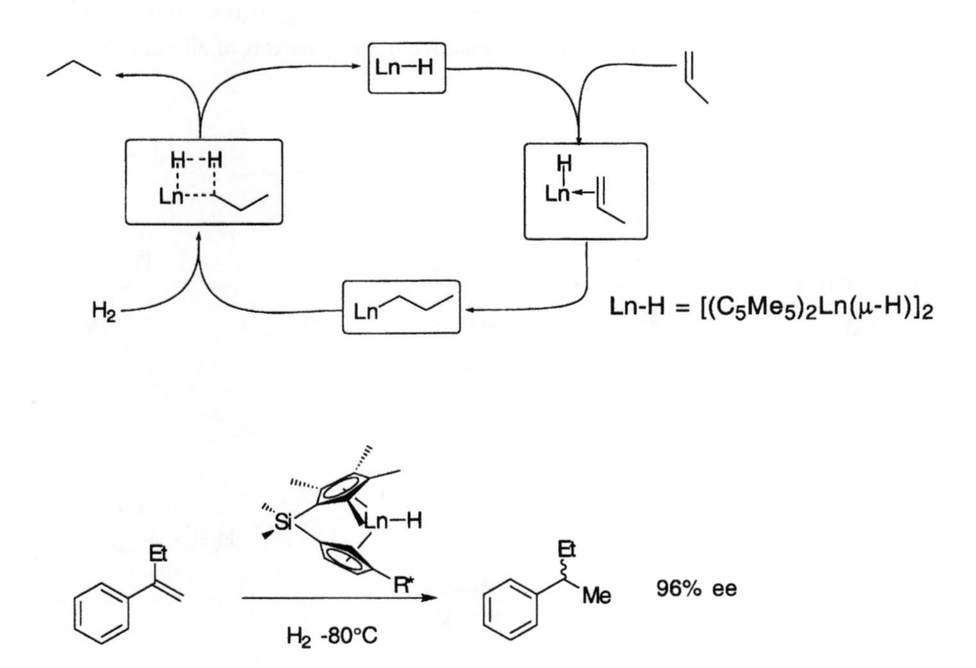

SCHEME 5.1 CATALYSIS OF ALKENE HYDROGENATION BY BIS(CYCLOPENTADIENYL)LANTHANIDE HYDRIDES.

The 'tied back' bridged cyclopentadienyl complexes $[(Me_2Si(C_5Me_4)_2)Ln(\mu\text{-}H)]_2$ show even higher activity for alkene hydrogenation because of less steric congestion at Ln. Enantioselective alkene hydrogenation can be achieved when a chiral group such as menthyl is introduced onto one of the cyclopentadienyl rings, and ee's of up to 96% have been achieved with a Sm catalyst by Giardello *et al.* (1994). The structure of the chiral pre-catalyst *R*-$[Me_2Si(Me_4C_5)(C_5H_3R^*)Ln\text{-}CH(SiMe_3)_2]$ with R* = (+)-neomenthyl is shown in Figure 5.2. Related chemistry of actinide complexes such as $[(Me_2Si(C_5Me_4)_2)ThH(\mu\text{-}H)]_2$ has also been explored by Fendrick *et al* (1984).

5.1.2 Hydroamination/Cyclization of Aminoalkenes and Aminoalkynes

The hydroamination/cyclization of unsaturated amines is a useful route to nitrogen-containing heterocycles. Marks and co-workers have made detailed studies of the catalysis of these reactions by organolanthanide complexes. This work is described by Giardello *et al.* (1994) and Li and Marks (1996). The precatalysts for these reactions are of the type $[Cp_2Ln\text{-}R]$ where R = H, alkyl, $N(SiMe_3)_2$, and the initiation step is the protonolysis of the Ln-R bond by the NH_2 group of the substrate. The next step of the reaction is an intramolecular alkene or alkyne insertion into the Ln-N bond, and the cyclized product is liberated by protonolysis of the Ln-C bond by another molecule of substrate. High enantioselectivities in the alkene hydroamination/cyclization have been achieved by using chiral 'tied-back' cyclopentadienyls like those used in enantioselective hydrogenation of alkenes (Section 5.1.1) as shown in Scheme 5.2.

Si Ln–R R*
R* = (-)-menthyl
Me Me H₂N
-HR
H Me Ln–N –Me
H N Ln Me Me
Me Me H₂N
H Me,,, N Me Me

H₂N R $(C_5Me_5)_2Ln\text{-}CH(SiMe_3)_2$ R H N R N

SCHEME 5.2 HYDROAMINATION/CYCLIZATION CATALYSED BY BIS(CYCLOPENTADIENYL) LANTHANIDE COMPLEXES.

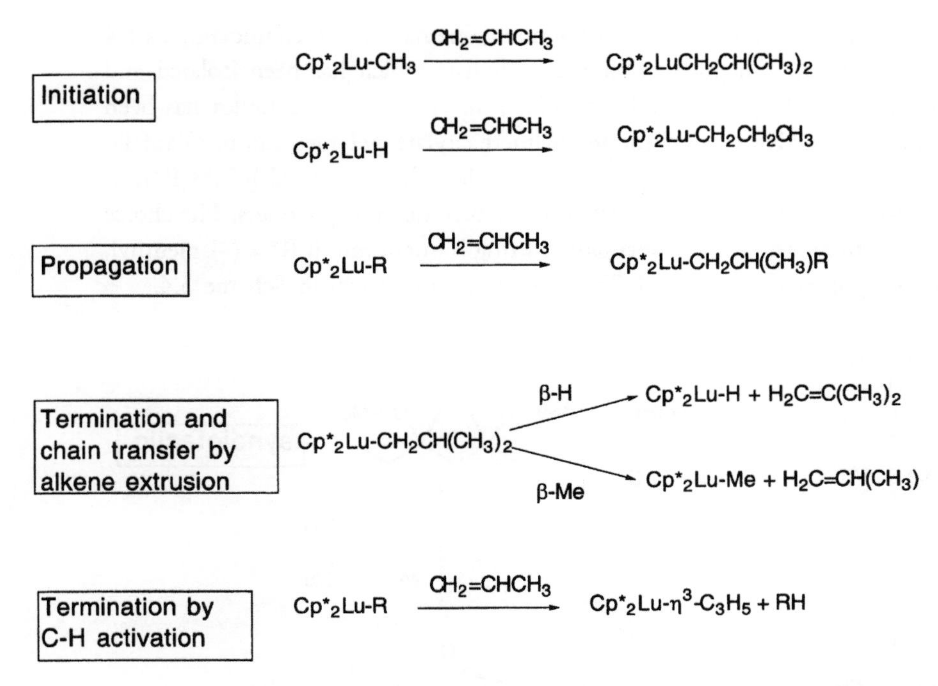

SCHEME 5.3 KEY STEPS IN ALKENE POLYMERIZATION CATALYZED BY $(C_5Me_5)_2LuH$.

5.1.3 Polymerization of Alkenes

Early investigations in this area were aimed at making models for Ziegler-Natta catalysis; the first alkene polymerization catalyzed by organolanthanides using $[(C_5H_4Me)_2Ln(\mu\text{-}Me)]_2$- was reported by Ballard *et al.* (1978). $[(C_5Me_5)_2LuMe]$, which exists in solution in equilibrium with $[(C_5Me_5)_2Lu(\mu\text{-}Me)LuMe(C_5Me_5)_2]$, was found by Watson and Parshall (1985) to be spectacularly active in alkene polymerization with turnovers greater than 1800 s^{-1} at 25°C and 1 atm of ethene. The key steps in the polymerization process have been elucidated in this model system and are summarized in Scheme 5.3.

The termination by β-Me elimination (the reverse of alkene insertion into the Lu-Me bond) is unique to lanthanide chemistry, and the termination by C-H activation led to the discovery of other C-H activation reactions of $[(C_5Me_5)_2LuMe]$ which are described in Chapter 4. As with alkene hydrogenation, increased activities are observed for the 'tied back' cyclopentadienyl complexes $[(Me_2Si(C_5Me_4)_2)Ln(\mu\text{-}H)]_2$ which have reduced steric hindrance at Ln.

5.1.4 Polymerization of Methacrylates

Bis(cyclopentadienyl)lanthanide hydrides and alkyls have also been applied to the commercially important polymerization of methylmethacrylate; these reactions are shown in Scheme 5.4. $[(C_5Me_5)_2Ln]$ derivatives give rise to highly syndiotactic living polymerization, producing high molecular weight polymers with very narrow polydispersity as described by Yasuda *et al.* (1993). The polymerization proceeds by an initial hydride attack on the CH_2 group of methylmethacrylate to give a transient enolate intermediate. Another molecule of methylmethacrylate then undergoes a 1,4 addition to give an 8-membered ring intermediate which has been isolated and characterized crystallographically by Yasuda *et al.* (1992). This complex has been shown to be an active catalyst for methylmethacrylate polymerization. Giardello *et al.* (1995) have used the analogous chiral complexes [$Me_2Si(Me_4C_5)(C_5H_3R^*)Ln$-$CH(SiMe_3)_2$] with R* = (+)-neomenthyl to produce isotactic polymers. The choice of chiral substituent on the cyclopentadienyl ring is important: if R* = (–)-menthyl, syndiotactic polymers are formed. These reactions are shown in Scheme 5.4.

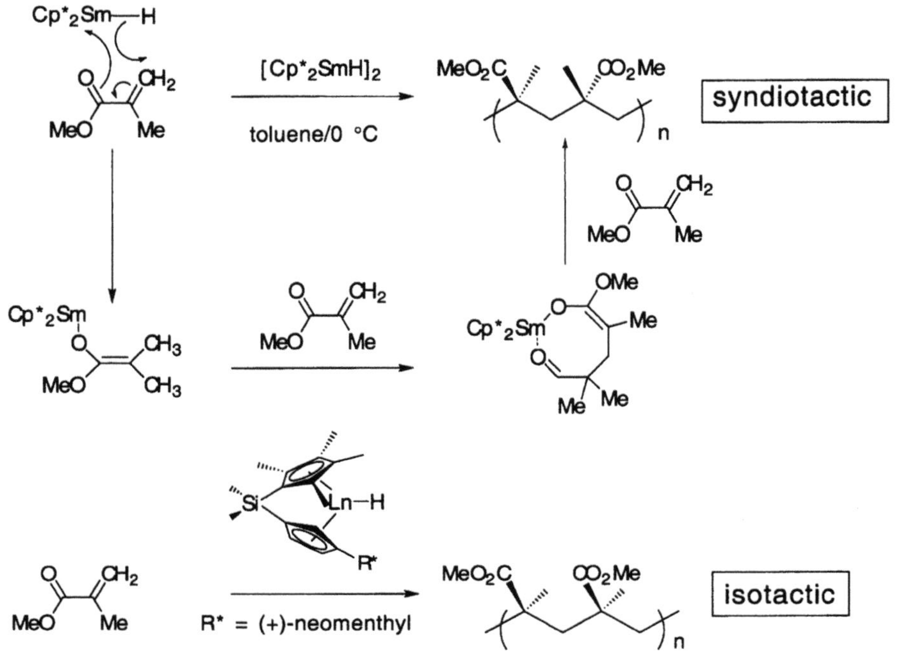

SCHEME 5.4 CATALYSIS OF METHACRYLATE POLYMERIZATION BY BIS(CYCLOPENTADIENYL) LANTHANIDE HYDRIDES.

- Labile substrate-to-Ln bonds allow high turnover rates in catalytic reactions.
- High catalytic activity requires readily available substrate binding sites at Ln.
- Catalytic activity and selectivity can be fine-tuned by varying Ln (change in ionic radius) or by varying the bulk of Cp ligands.
- 'Tied-back' bis(cyclopentadienyl) ligands allow easy access of substrate to Ln.
- Enantioselective catalysis can be achieved by introducing chiral substituents onto Cp ligands
- Important catalytic reactions of bis(cyclopentadienyl) lanthanide complexes include alkene hydrogenation, hydroamination and polymerization.

5.2 LEWIS ACID CATALYSIS BY LANTHANIDE SALTS AND COMPLEXES

Activation of substrates by coordination to a Lewis acid is crucial in many important organic transformations such as Friedel-Crafts and Diels-Alder reactions. As well as activation of substrates, Lewis acids may also contribute to selectivity by organization of coordinated substrates in the transition state, and there are a growing number of examples of enantioselective Lewis acid catalysis. 'Classical' Lewis acids such as BF_3, $AlCl_3$ and $TiCl_4$ may be effective at promoting reactions but their moisture sensitivity, requiring handling under strictly anhydrous conditions, leads to handling difficulties, and their relatively strong metal-to-ligand bonds mean that dissociation of product after reaction may be very slow, and thus stoichiometric quantities of the Lewis acid may be required. Compared with these classical Lewis acids, lanthanide salts have several attractive features:

- They are typical 'hard' Lewis acids.
- Lanthanide-to-ligand bond is usually labile, leading to rapid dissociation of product and thus the possibility of high catalytic turnover.
- Large ionic radius and flexible coordination geometry allows binding of a wide range of substrates.

The Lewis acidity of lanthanide salts increases with decreasing ionic radius and therefore an increase in activity is generally observed on traversing the series from La to Lu, and Sc, although expensive, is often used when particularly high activity is required. Because of its high cost, Lu is almost never used; the next smallest lanthanide Yb is usually chosen because of its reasonably low price and generally high activity.

5.2.1 Lanthanide tris(β-diketonates)

The lanthanide tris(β-diketonates) [Ln(fod)$_3$] (fod = heptafluorooctanedionate), better known as NMR shift reagents (see Chapter 2), were among the first lanthanide complexes to be used as Lewis acid catalysts. They are very soluble in organic solvents and their mild Lewis acidity means that they may be used with acid sensitive substrates. Both [Eu(fod)$_3$] and [Yb(fod)$_3$] have been applied with great success to hetero Diels-Alder reactions where the conventional Lewis acid $ZnCl_2$ is required in stoichiometric quantities. Excellent regio- and diastereoselectivity is achieved as shown in Scheme 5.5.

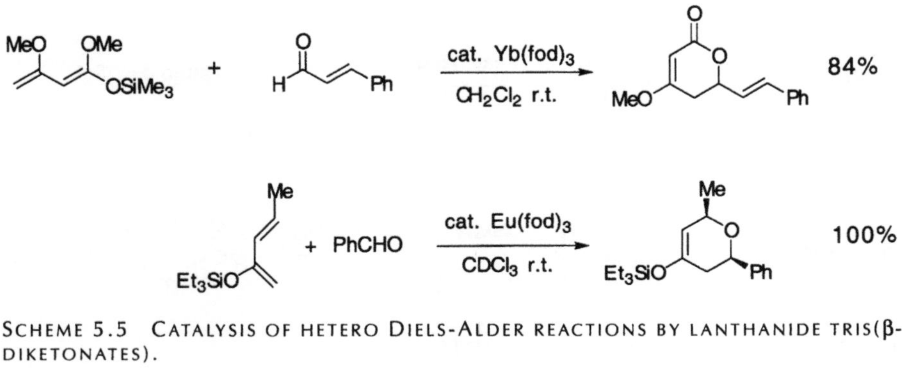

SCHEME 5.5 CATALYSIS OF HETERO DIELS-ALDER REACTIONS BY LANTHANIDE TRIS(β-DIKETONATES).

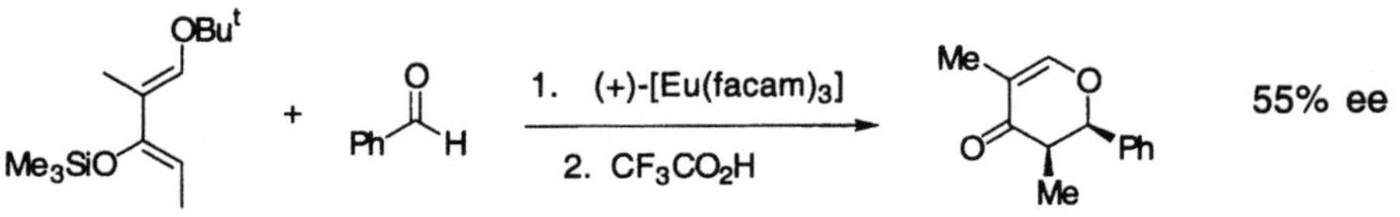

SCHEME 5.6 ENANTIOSELECTIVE HETERO DIELS-ALDER REACTION CATALYZED BY A CHIRAL LANTHANIDE COMPLEX.

Enantioselectivity can also be achieved by using the chiral lanthanide shift reagent [Eu(facam)$_3$] (facam = (3-heptafluoropropylhydroxymethylene) camphorato) as shown in Scheme 5.6.

5.2.2 Lanthanide Triflates

The most successful rare earth Lewis acid catalysts have been the triflate salts, Ln(OTf)$_3$ (OTf = $CF_3SO_3^-$). Their applications to organic synthesis have been reviewed by Marshman (1995) and by Kobayashi (1994). The highly electron-withdrawing nature of the OTf group means that these salts are stable to hydrolysis; in the presence of water they exist as hydrates [Ln(H_2O)$_9$][OTf]$_3$, but they can be dehydrated by prolonged heating under vacuum to give the anhydrous salts. Anhydrous Ln(OTf)$_3$ are sparingly soluble in organic solvents such as CH_2Cl_2 and have been used to catalyze a range of reactions summarized in Scheme 5.7. On aqueous work-up the catalysts are extracted into the aqueous layer and can be re-used after drying without significant loss of activity. Water is the ultimate environmentally friendly solvent for organic synthesis and there are examples of rare-earth (and scandium) triflate catalysis in H_2O and in mixed H_2O/THF solvent.

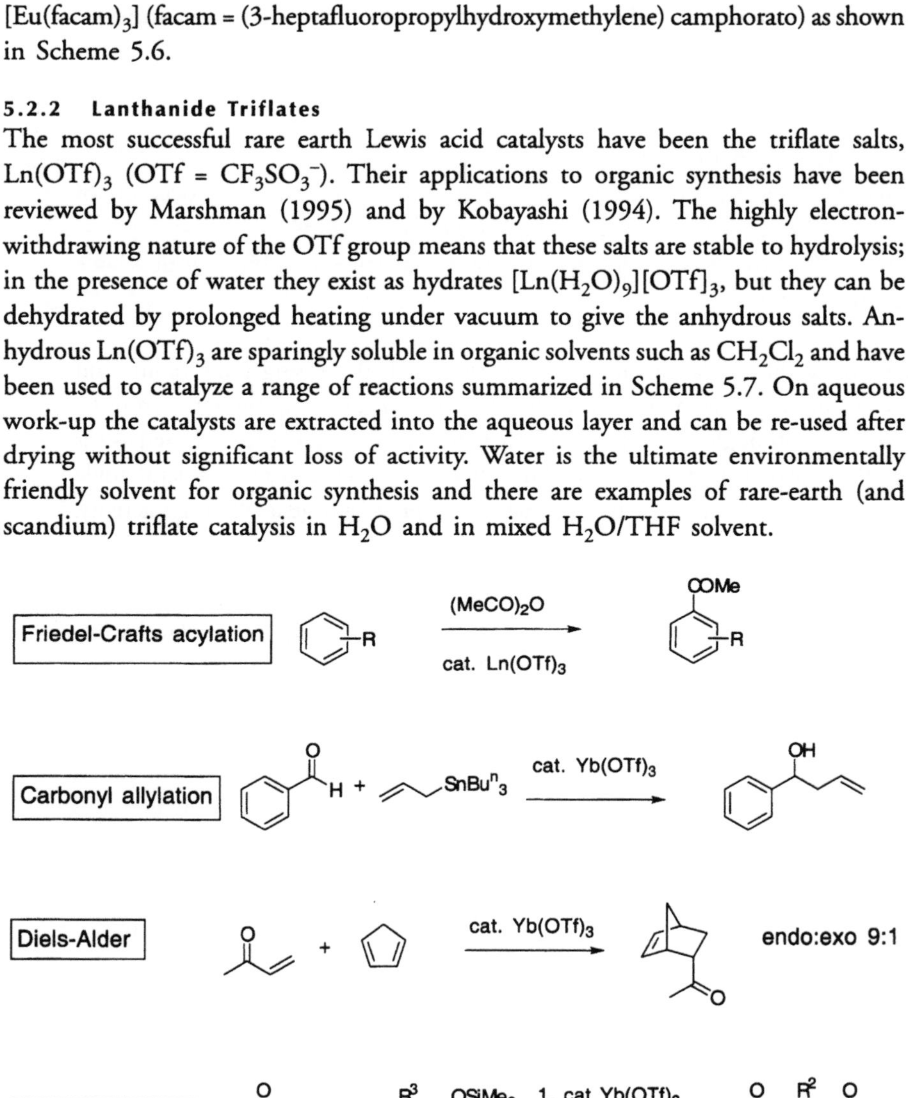

SCHEME 5.7 LEWIS ACID CATALYSIS BY Ln(OTf)$_3$.

Enantioselective catalysis by lanthanide triflates

Highly enantioselective catalysts can be made by addition of chiral auxilliaries to $Ln(OTf)_3$. Kobayashi (1994) has added the chiral diol binaphthol together with a bulky amine to $Yb(OTf)_3$ to produce an enantioselective catalyst for the Diels Alder reaction as shown in Scheme 5.8. The structure of the catalyst has been tentatively assigned by solution spectroscopy, but the complex has not been isolated and characterized crystallographically.

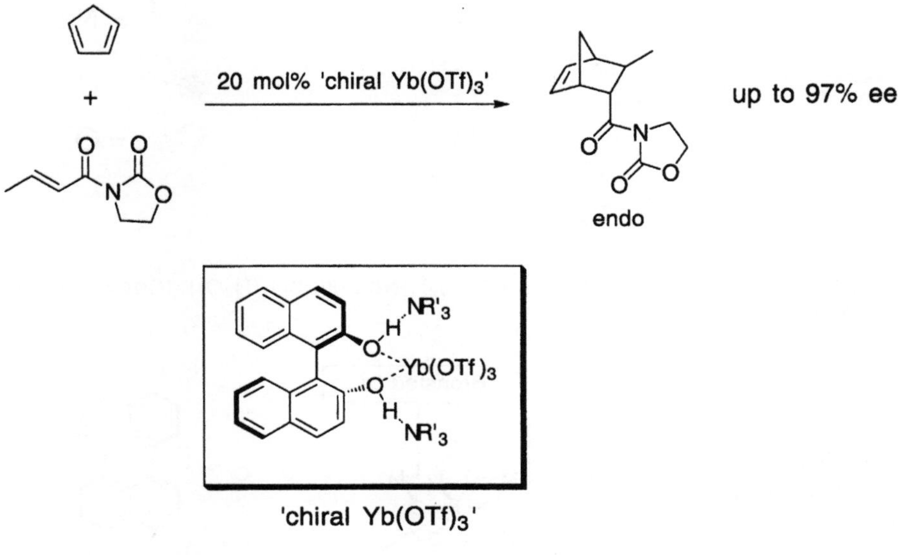

20 mol% 'chiral $Yb(OTf)_3$'

up to 97% ee

endo

'chiral $Yb(OTf)_3$'

SCHEME 5.8 ENANTIOSELECTIVE DIELS-ALDER REACTION CATALYZED BY A CHIRAL YB TRIFLATE.

5.2.3 Alkali metal lanthanide (tris)binaphtholates

These heterometallic complexes show catalytic activity as both Lewis acids and Brønsted bases, and have been applied to a range of enantioselective reactions as described by Shibasaki, Sasai and Arai (1997). The complexes are prepared by reaction of $LaCl_3$ with Li_2binol (H_2binol = binaphthol) and $NaOBu^t$ in the presence of H_2O, and were originally believed to be monomeric hydroxides of formula [(binol)LaOH]. However they were subsequently found by a combination of mass spectrometry and X-ray diffraction to be heterometallic complexes of general formula $[M(THF)_2]_3[Ln(binol)_3(H_2O)]$, the crystal structure of which is shown in Figure 5.3. They can act as Brønsted bases by virtue of the metal alkoxide functionality, and the Ln ion can act as a Lewis acid.

The first highly enantioselective reaction catalyzed by these complexes was the nitroaldol reaction, where Brønsted base reactivity leads to formation of a nitronate complex. The carbonyl group of the aldehyde can coordinate to the Lewis acidic Ln centre and enantioselective C-C bond formation can occur. The enantioselective hydrophosphonylation of aldehydes is another example of bifunctional catalysis by $[M(THF)_2]_3[Ln(binol)_3(H_2O)]$. Scheme 5.9 shows the enantioselective nitroaldol

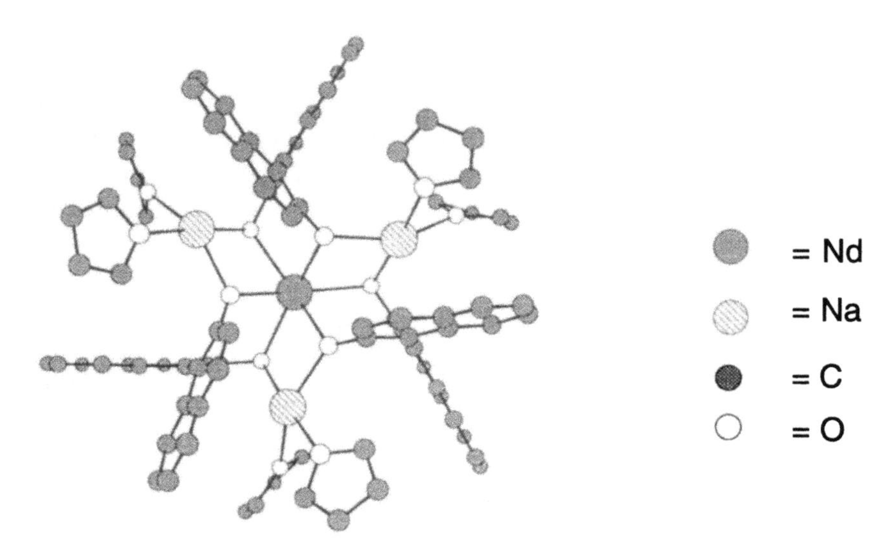

FIGURE 5.3 CRYSTAL STRUCTURE OF $[Na(THF)_2]_3[Nd(S\text{-BINOL})_3(H_2O)]$. (COORDINATED H_2O IS HIDDEN BY ND).

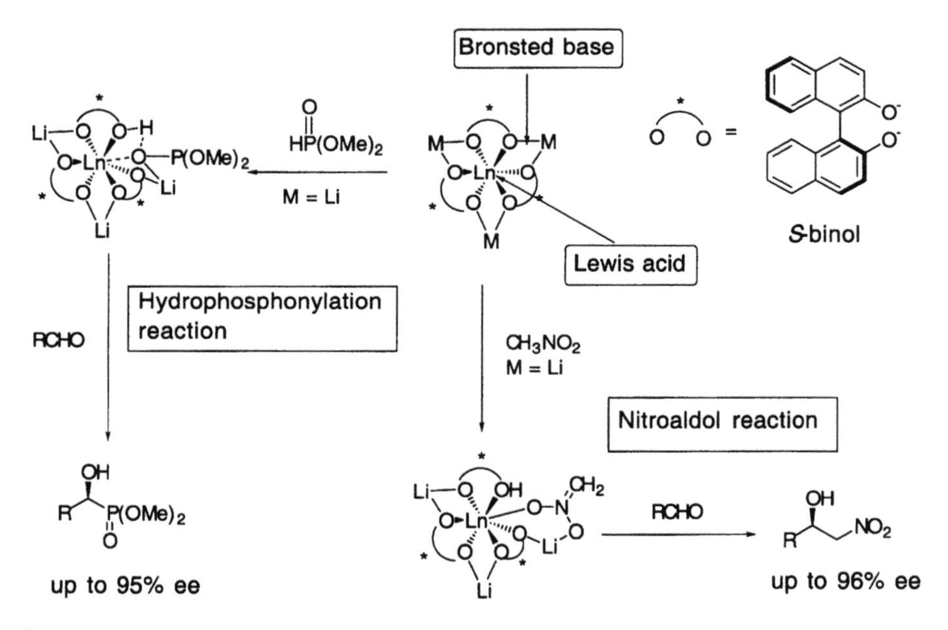

SCHEME 5.9 ENANTIOSELECTIVE NITROALDOL AND HYDROPHOSPHONYLATION REACTIONS CATALYZED BY $M_3[Ln(BINOL)_3]$.

and hydrophosphonylation reactions. Enantioselective Lewis acid catalysis of Michael and Diels-Alder reactions by $[M(THF)_2]_3[Ln(binol)_3(H_2O)]$ have also been reported. The enantioselectivity of these catalysts is sensitive to both M and Ln: the nitroaldol reaction is most effective when M=Li and Ln=Eu whereas the Michael reaction requires M=Na and Ln=La. In contrast to Lewis acid catalysis by $Ln(OTf)_3$, the heterometallic catalysts are least effective with small later lanthanides.

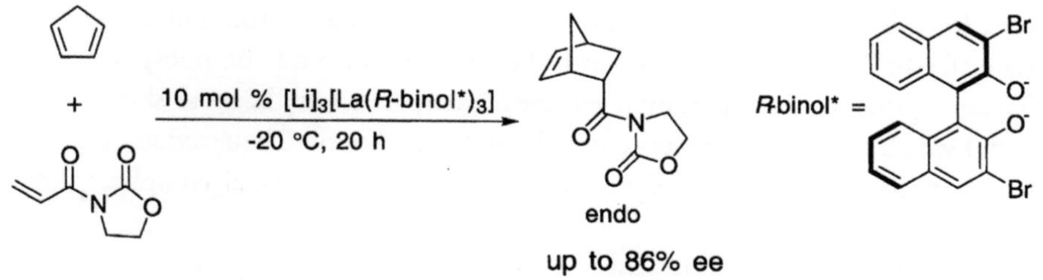

SCHEME 5.10 ENANTIOSELECTIVE DIELS-ALDER REACTION CATALYZED BY $M_3[Ln(BINOL)_3]$.

$[M(THF)_2]_3[Ln(binol)_3(H_2O)]$ can also act as effective enantioselective Lewis acid catalysts and ee's of up to 86% have been achieved by Morita *et al.* (1998) in the Diels-Alder reaction as shown in Scheme 5.10. The Li cations are believed to be important in activation of the dienophile as addition of 12-crown-4, a Li^+ complexing agent, leads to much lower ee's.

5.2.4 Nucleic acid hydrolysis

A range of lanthanide complexes have now been investigated as catalysts for the hydrolytic cleavage of RNA as reviewed by Morrow *et al.* (1995). This reaction proceeds *via* coordination of the phosphate group to the metal centre followed by nucleophilic attack by OH^- as shown in Figure 5.4.

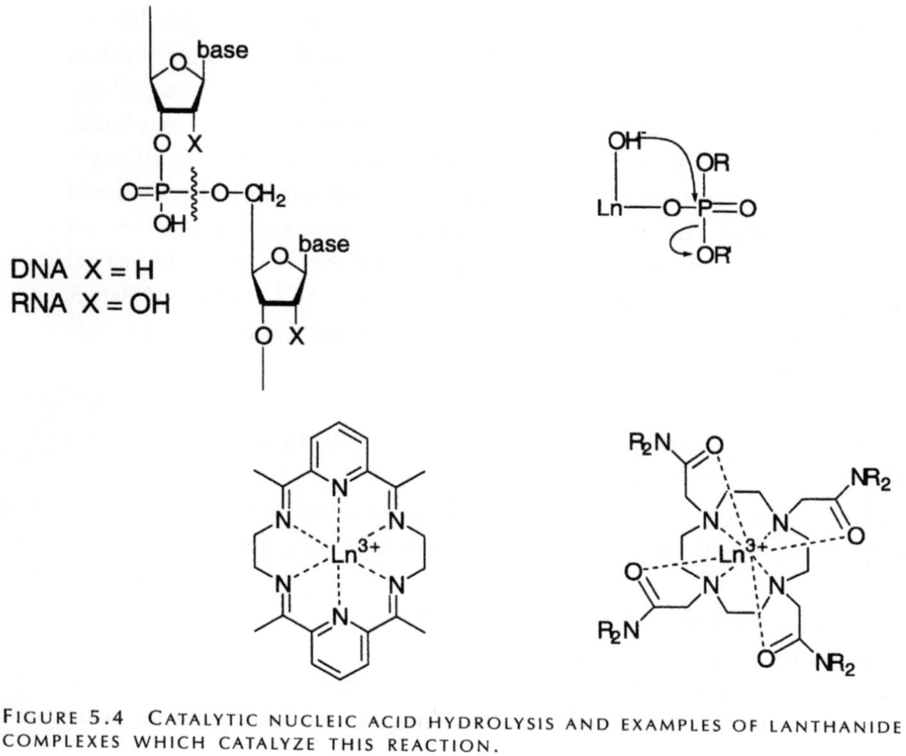

FIGURE 5.4 CATALYTIC NUCLEIC ACID HYDROLYSIS AND EXAMPLES OF LANTHANIDE COMPLEXES WHICH CATALYZE THIS REACTION.

Catalysts for this reaction must be stable and soluble in aqueous solution at the appropriate pH (≥ 7) and must be suffiently Lewis acidic to bind the phosphate group. These requirements are met by macrocyclic complexes such as those shown in Figure 5.4 which catalyze RNA cleavage under mild conditions. The reaction can be made site-specific by attaching oligonucleotide groups to the metal complex.

- Ln^{3+} ions are typical hard Lewis acids.
- Labile Ln-to-substrate bonds allow rapid turnover rates.
- Reactions catalyzed include: Friedel-Crafts acylation, carbonyl allylation, Diels-Alder and nucleic acid hydrolysis.
- Enantioselective catalysis can sometimes be achieved by use of well characterized chiral complexes or by addition of chiral ligands.

5.3 ORGANOCERIUM REAGENTS

The use of organocerium reagents in organic synthesis was first reported by Imamoto *et al.* (1984). Cerium was chosen simply on the grounds of its low cost (there is no significant difference in activity on traversing the whole lanthanide series), and the reagents are prepared by the reaction of anhydrous $CeCl_3$ or CeI_3 with the appropriate LiR in THF at –78°C. The reagents are used at this temperature; warming to 0°C results in decomposition if the R group contains a β-hydrogen, and reactions at higher temperatures often result in mixtures of products sometimes due to reduction reactions. Organoceriums are particularly useful for addition to carbonyl groups, reacting much more cleanly and selectively than either organolithium or Grignard reagents. They have reduced basicity compared with organolithium reagents and so can be used for additions to enolizable carbonyls. The high oxophilicity of cerium leads to selective addition to C=O in the presence of C-halogen bonds in the substrate, and also results in selective addition to C=O in unsaturated carbonyls. Scheme 5.11 summarizes representative reactions of organoceriums which are useful in synthetic chemistry. The exact nature of organoceriums is not known; they are unlikely to be simple LnR_3 and it has been proposed that they may have the general formula $Li_3[CeR_3X_3]$. The hexamethyl complexes $[Li(TMEDA)]_3[LnMe_6]$ (Chapter 4), which are stable at room temperature, show similarly selective reactivity towards unsaturated carbonyl compounds.

- Reduced basicity of organocerium reagents compared with organolithium reagents allows addition to enolizable carbonyls.
- Oxophilicity and mild Lewis acidity of Ce allows organocerium reagents to add selectively to C=O in the presence of C-halogen and C=C bonds.

5.4 CATALYSIS BY LANTHANIDE ALKOXIDES

5.4.1 Meerwein-Ponndorf-Verley Reduction

The Meerwein-Ponndorf-Verley reduction is an equilibration of 2° alcohols with ketones resulting in reduction of the ketone. This reaction can be mediated by

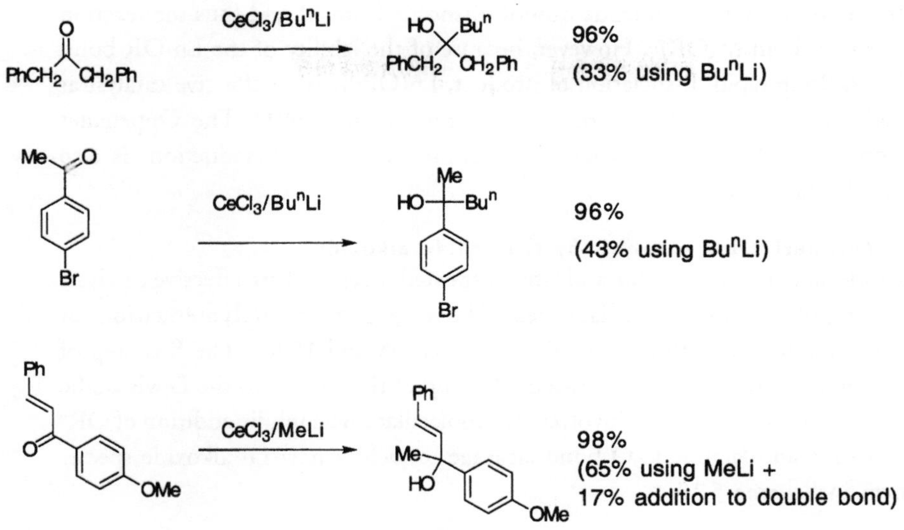

SCHEME 5.11 ADDITION OF ORGANOCERIUM REAGENTS TO CARBONYL COMPOUNDS.

stoichiometric quantities of metal alkoxides such as $Al(OPr^i)_3$, but can be catalyzed by as little as 5 mol% of $Ln(OR)_3$. The reaction proceeds *via* coordination of the ketone to the Lewis acidic metal centre, followed by hydride transfer from the alkoxide to the ketone as outlined in Scheme 5.12. If the M-OR bond is strong,

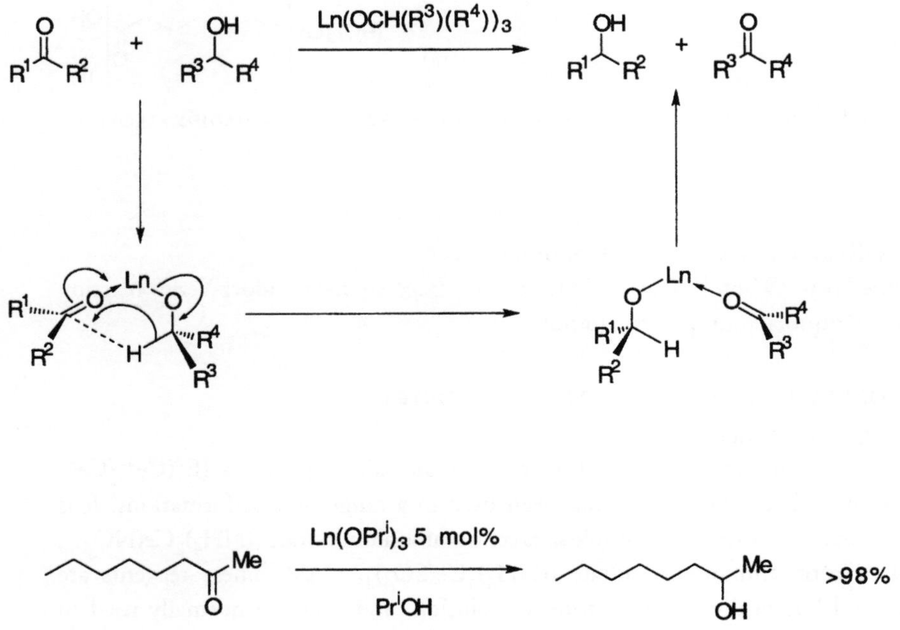

SCHEME 5.12 MEERWEIN-PONNDORF-VERLEY REDUCTIONS CATALYZED BY LANTHANIDE ALKOXIDES.

then the newly-formed 2° alkoxide cannot dissociate from M and thus the reaction is stoichiometric in $M(OR)_3$. However, because of the lability of the Ln-OR bond and the resulting rapid dissociation of product, $Ln(OPr^i)_3$ is an effective catalyst at low loadings as reported by Lebrun, Nancy and Kagan (1991). The Oppenauer oxidation of 2° alcohols to ketones, the reverse of the MVP reduction, is also catalyzed by $Ln(OR)_3$.

5.4.2 Polymerization catalysis by lanthanide alkoxides

Lanthanide alkoxides are soluble and labile Brønsted bases, and are effective catalysts for living polymerization of lactones. The ring-opening polymerization of ε-caprolactone has been studied by Shen *et al.* (1995 and 1996). The first step of the reaction is coordination of the carbonyl group of the lactone to the Lewis acidic Ln atom. The propagation step involves intramolecular nucleophilic addition of OR^- to the lactone and lactone C-O bond cleavage to yield a new Ln alkoxide species as shown in Scheme 5.13.

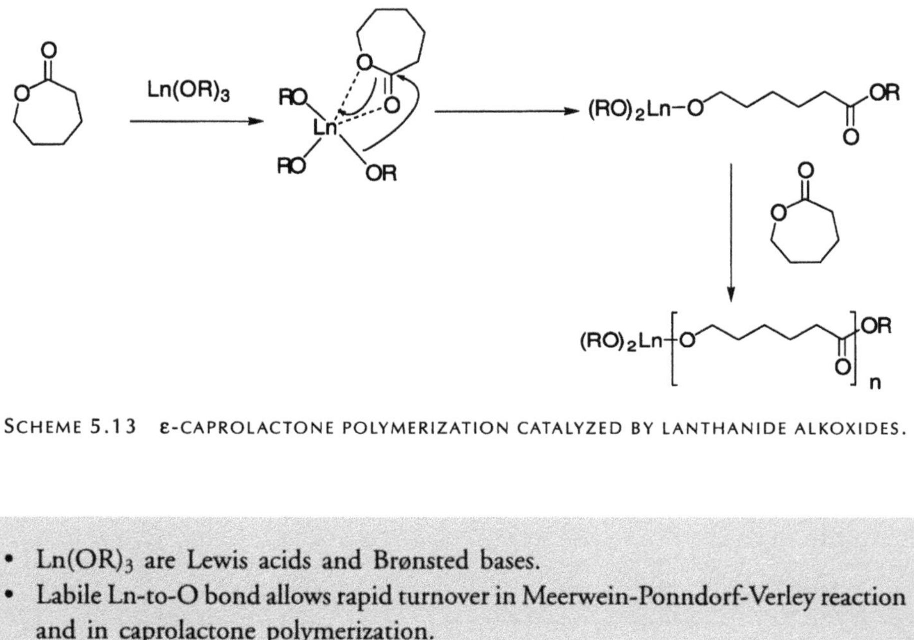

Ln(OR)3

RO RO Ln O O OR

(RO)2Ln–O OR O

(RO)2Ln O OR O n

SCHEME 5.13 ε-CAPROLACTONE POLYMERIZATION CATALYZED BY LANTHANIDE ALKOXIDES.

- $Ln(OR)_3$ are Lewis acids and Brønsted bases.
- Labile Ln-to-O bond allows rapid turnover in Meerwein-Ponndorf-Verley reaction and in caprolactone polymerization.

5.5 OXIDATION AND REDUCTION CHEMISTRY

5.5.1 Ce(IV) Oxidations

Ce(IV) is one of the most powerful one-electron oxidizing agents ($E°(Ce^{4+}/Ce^{3+}$ $=1.72V$ in acid solution) and it has been used in a range of transformations. It is usually used in the form of the double salts ceric ammonium nitrate ($(NH_4)_2Ce(NO_3)_6$, CAN) or ceric ammonium sulfate ($(NH_4)_4Ce(SO_4)_4$, CAS). These reagents are soluble and kinetically stable in aqueous solution, and they are normally used in mixed aqueous/organic solvent systems such as H_2O/MeCN. Important transformations include oxidation of polyaromatic systems to quinones, selective side chain

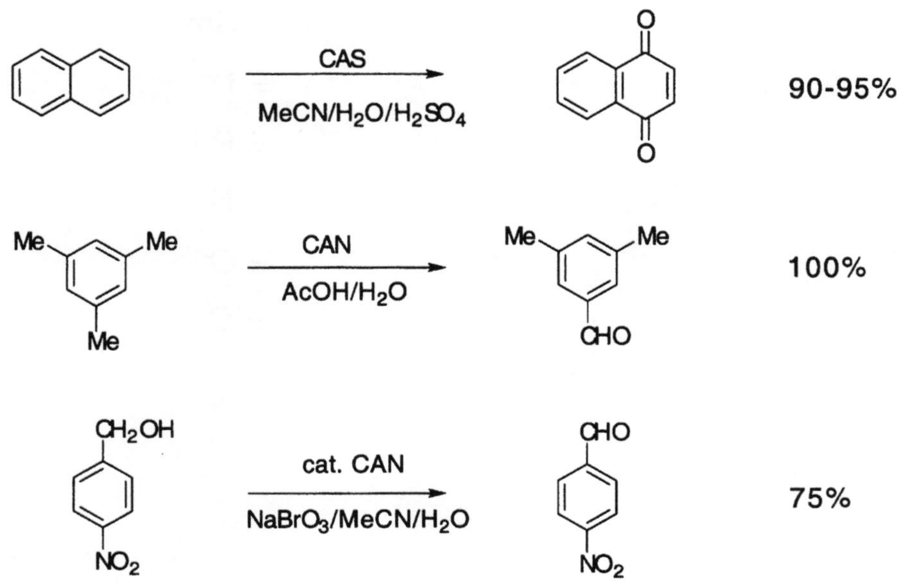

CAS
$MeCN/H_2O/H_2SO_4$
90-95%

Me Me Me
CAN
$AcOH/H_2O$
Me Me CHO
100%

CH_2OH NO_2
cat. CAN
$NaBrO_3/MeCN/H_2O$
CHO NO_2
75%

SCHEME 5.14 OXIDATIONS BY Ce(IV) COMPOUNDS.

oxidation of arenes, and oxidation of alcohols to aldehydes as shown in Scheme 5.14. Ce(IV) can also be used as a catalyst for stoichiometric oxidations by $NaBrO_3$.

5.5.2 Samarium(II) Iodide Reductions

SmI_2 in THF has a useful E° (–1.33 V) for use in organic synthesis; a convenient preparation of SmI_2 was first reported by Kagan in 1977, and since then it has become a valuable one-electron reducing agent in organic chemistry. SmI_2 can be prepared as a deep blue solution in THF by direct reaction of Sm metal with 1,2-diiodoethane as shown below:

$$Sm + ICH_2CH_2I \xrightarrow{THF} SmI_2 + H_2C = CH_2$$

Its applications have recently been reviewed by Molander and Harris (1996) and by Krief and Laval (1999). It is a selective inner-sphere electron-transfer reagent and has the advantages of being reasonably soluble in THF and being a mild Lewis acid, thus allowing chelation control in some of its reactions. Its reduction potential can be increased by addition of strong donor ligands such as HMPA (HMPA = $(Me_2N)_3PO$) which forms the six-coordinate adduct $[SmI_2(HMPA)_4]$ as shown in Figure 5.5.

Some examples of SmI_2 mediated reactions are shown in Scheme 5.15; the high stereoselectivity shown in the intramolecular Reformatsky reaction is due to chelation control by the Lewis acidic Sm centre. Successful use of catalytic quantities of SmI_2 in the presence of Zn amalgam has recently been reported by Corey and Zheng (1997), making SmI_2 reductions even more attractive.

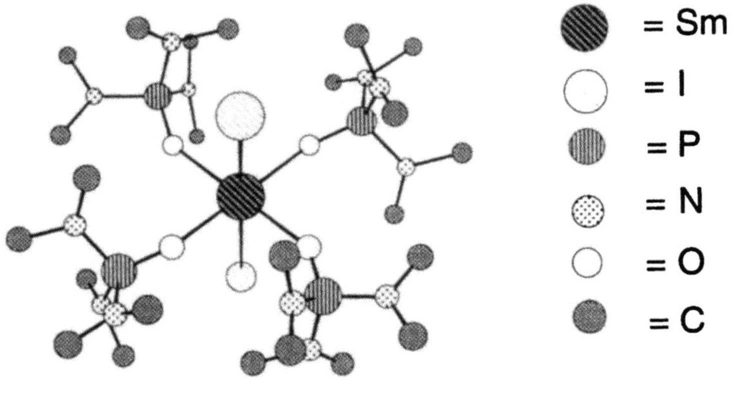

FIGURE 5.5 STRUCTURE OF $[SmI_2(HMPA)_4]$.

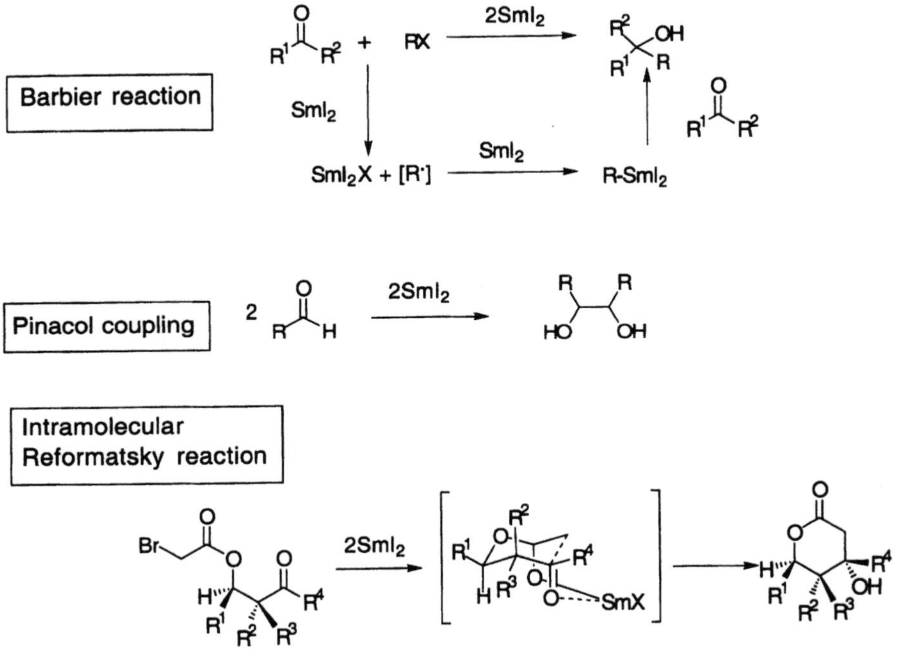

SCHEME 5.15 SmI_2 MEDIATED REDUCTIONS.

- Ce(IV) ($4f^0$) is a powerful 1-electron oxidizing agent.
- Sm(II) ($4f^6$) is a useful 1-electron reducing agent.
- Mild Lewis acidity of Sm allows stereoselectivity by chelation control for some Sm(II) reductions.

REFERENCES

Ballard, D.G.H., Courtis, A., Holton, J., McMeeking, J. and Pearce, R.(1978) Alkyl Bridged Complexes of THE Group 3A and Lanthanoid Metals as Homogeneous Ethylene Polymerisation Catalysts. *J. Chem. Soc., Chem. Commun.*, 994–995.

Corey, E.J. and Zheng, G. Z. (1997) Catalytic Reactions of Samarium (II) Iodide. *Tetrahedron Lett.*, **38**, 2045–2048.

Fendrick, C.M., Mintz, E.A., Schertz, L.D., Marks, T.J. and Day, V.W. (1984) Manipulation of Organoactinide Coordinative Unsaturation and Stereochemistry. Properties of Chelating Bis(Polymethylcyclopentadienyl) Hydrocarbyls and Hydrides.*Organometallics*, **3**, 819–821.

Giardello, M.A., Conticello, V.P., Brard, L., Sabat, M., Rheingold, A.L., Stern, C.L. and Marks, T.J. (1994a) Chiral organolanthanides designed for asymmetric catalysis. Synthesis, characterization, and configurational interconversions of chiral, C_1-symmetric organolanthanide halides, amides and hydrocarbyls. *J. Am. Chem. Soc.*, **116**, 10212–10240.

Giardello, M.A., Conticello, V.P., Brard, L., Gagné, M.R. and Marks, T.J. (1994b) Chiral Organolanthanides Designed for Asymmetric Catalysis. A Kinetic and Mechanistic Study of Enantioselective Hydroamination/Cyclization and Hydrogenation by C_1-Symmetric $Me_2Si(Me_4C_5)(C_5H_3R^*)Ln$ Complexes where R^* = Chiral Auxilliary. *J. Amer. Chem. Soc.*, **116**, 10241–10254.

Giardello, M.A., Yamamoto, Y., Brard, L., Marks, T.J. (1995) Stereocontrol in the Polymerization of Methyl Methacrylate Mediated by Chiral Organolanthanide Metallocenes. *J. Am. Chem. Soc.*, **117**, 3276–3277.

Imamoto, T. (1994) *Lanthanides in Organic Synthesis*, London: Academic Press.

Imamoto, T., Kusumoto, Y., Tawarayama, Y., Sugiura, Y., Mita, T., Hatanaka, Y., Yokoyama, M. (1984) Carbon-Carbon Bond-Forming Reactions Using Cerium Metal or Organocerium Reagents *J. Org. Chem.*, **49**, 3904–3912.

Kobayashi, S. (1994) Rare Earth Metal Trifluoromethanesulfonates as Water-Tolerant Lewis Acid Catalysts. *Synlett*, 689–701.

Kobayashi, S. (ed) (1999) *Lanthanides: Chemistry and Use in Organic Synthesis*, Berlin Heidelberg: Springer-Verlag.

Kobayashi, S., Ishitani, H. (1994) Lanthanide(III)-Catalysed Enantioselective Diels-Alder Reactions. Stereoselective Synthesis of Both Enantiomers by Using a Single Chiral Source and a Choice of Achiral Ligands. *J. Am. Chem. Soc.*, **116**, 4083–4084.

Krief, A., Laval, A.-M. (1999) Coupling of Organic Halides with Carbonyl Compounds Promoted by SmI_2, the Kagan Reagent. *Chem. Rev.*, **99**, 745–777.

Lebrun, A., Namy, J.L. and Kagan, H.B. (1991) A New Preparation of Lanthanide Alkoxides and some Applications in Catalysis. *Tetrahedron Lett.*, **32**, 2355.

Li, Y.W. and Marks, T.J. (1996) Organolanthanide-catalyzed intramolecular hydroamination/cyclization of aminoalkynes. *J. Amer. Chem, Soc.*, **118**, 9295–9306.

Marshman, R.W. (1995) Rare-Earth Triflates in Organic Synthesis. *Aldrichimica Acta*, **28**, 77–84.

Molander, G.A. and Harris, C.R. (1996) Sequencing Reactions with Samarium (II) Iodide. *Chem. Rev.*, **96**, 307–338.

Morita, T., Arai, T., Sasai, H. and Shibasaki, M. (1998) Utilization of heterobimetallic complexes as Lewis acids. *Tetrahedron-Asymmetry*, **9**, 1445–1450.

Morrow, J.R., Kolasa, K.A., Amin, S. and Aileen, K.O. (1995) Metal ion macrocyclic complexes as artificial ribonucleases. *Advances In Chemistry Series*, **246**, 431-447.

Shen, Y., Shen, Z., Shen, J., Zhang, Y. and Yao, K. (1996) Characteristics and mechanism of ε-caprolactone polymerization with rare earth halide systems. *Macromolecules*, **29**, 3441–3446.

Shen, Y., Shen, Z., Zhang, F. and Zhang, Y. (1995) Ring opening polymerization of ε-caprolactone by rare earth alkoxide-CCl_4 system. *Polymer Journal*, **27**, 59–64.

Shibasaki, M., Sasai, H. and Arai, T. (1997) Asymmetric Catalysis with Heterobimetallic Compounds. *Angew. Chem. Int. Ed.*, **36**, 1236–1256.

Watson, P.L. and Parshall, G.W. (1985) Organolanthanides in Catalysis. *Acc. Chem. Res.*, **18**, 51–56.

Yasuda, H. and Ihara, E. (1997) Living polymerizations of polar and nonpolar monomers by the catalysis of organo rare earth metal complexes. *Bull. Chem. Soc. Japan*, **70**, 1745–1767.

Yasuda, H., Yamamoto, H., Yamashita, M., Yokota, K., Nakamura, A., Miyake, S., Kai, Y. and Kanehisa, N. (1993) Synthesis of High Molecular Weight Poly(methyl methacrylate) with Extremely Low Polydispersity by the Unique Function of Organolanthanide(III) Complexes. *Macromolecules*, **26**, 7134–7143.

INDEX

For Product Safety Concerns and Information please contact our EU representative GPSR@taylorandfrancis.com
Taylor & Francis Verlag GmbH, Kaufingerstraße 24, 80331 München, Germany

www.ingramcontent.com/pod-product-compliance
Ingram Content Group UK Ltd.
Pitfield, Milton Keynes, MK11 3LW, UK
UKHW051827180425
457613UK00007B/238